MEMOIRS
OF THE
AMERICAN MATHEMATICAL SOCIETY

Number 112

ON THE MIXED PROBLEM FOR A HYPERBOLIC EQUATION

by

TADEUSZ BAŁABAN

Published by the
American Mathematical Society
Providence, Rhode Island
1971

On the Mixed Problem for a Hyperbolic Equation

T. Bałaban
Institute of Mathematics, University, Warsaw

0. Introduction

The aim of this paper is to present the existence theorems for the mixed problem for certain class of hyperbolic operators with boundary conditions. The subject was stimulated by S. Agmon's results announced in [1]. He has considered the operators with constant coefficients in the main part of operators and in domains bounded by suitable hyperplanes. We generalize his results for operators with variable coefficients and for domains bounded by hypersurfaces.

Because the theory of the Cauchy problem for hyperbolic operators is well developed (see [3], [4]), it is sufficient to consider the boundary problem only. Also it is sufficient to consider this problem locally, e.g. to seek the solutions in a neighbourhood of the boundary hypersurface.

As usually the existence theorem is reduced to some inequality, so called basic inequality, expressing the continuity of the inverse operator in a suitable function space. To formulate this inequality let us introduce a few definitions. Let $\Omega \subset R^{n+1}$ be a domain, φ , $\psi \in C^{\infty}(\Omega)$ be a real functions with $\operatorname{grad}\varphi(x) \neq 0$ for $x \in \Omega$, $\operatorname{grad}\psi(x) \neq 0$ when $\psi(x) = 0$, and let $P(x,D)$ be a differential operator

Received by the editors September 15, 1970.

of order m defined on Ω, strongly hyperbolic with respect to grad $\varphi(x)$ and such that the hypersurface $\psi(x) = 0$ is non-characteristic for it.

Assume that on the hypersurface $\psi(x) = 0$ there are defined a differential operators $Q_j(x,D)$ of order m_j, $j = 1, \ldots, \varkappa$.

Let us denote $\Omega^+ = \{x \in \Omega : \psi(x) > 0\}$, $S = \{x \in \Omega : \psi(x) = 0\}$.

The results of Agmon ([1]) and the inequalities obtained for the problems in two independent variables (see [8], [9]) and for the operators of second order (see [5], [6]) suggest that the following inequality

$$(0.1) \qquad |\eta_0| \sum_{|\alpha|<m} \int_{\Omega^+} |D^\alpha u|^2 e^{2\eta_0\varphi} dx + \sum_{|\alpha|<m} \int_S |D^\alpha u|^2 e^{2\eta_0\varphi} dS \leq$$

$$\leq K\left(\int_{\Omega^+} |P(x,D)u|^2 e^{2\eta_0\varphi} dx + \sum_{j=1}^{\varkappa} \sum_{|\alpha|<m-m_j} \int_S |D^\alpha Q_j(x,D)u|^2 e^{2\eta_0\varphi} dS \right)$$

holds for $u \in C_0^\infty(\Omega)$, $|\eta_0|$ sufficiently large, with a constant K independent of u and η_0.

We shall prove this inequality under the same assumptions as in the paper of Agmon. The fourth section of this paper is devoted to an analysis of these assumptions. We discuss them from the point of view of wishful theory of the mixed problem.

The paper is divided into four sections. The first section contains the proof of the basic inequality. The second contains the proof of the dual inequality.

The existence theory of the mixed problem on the manifolds is the subject of the third section.

The main results of this paper without proofs were already published in the note [2].

We shall gather here denotations used in all sections.

$x'' = (x_1,\dots,x_{n-1}) \in R^{n-1}$, $x' = (x_0,x'') \in R^n$,

$x = (x',x_n) \in R^{n+1}$, similarly ξ'', ξ' and ξ,

$\zeta' = (\zeta_0, \xi'')$, $\zeta_0 = \xi_0 + i\eta_0$, $\zeta = (\zeta', \xi_n)$,

$\eta = (\eta_0, 0,\dots,0)$;

$E_+^{n+1} = \{x \in R^{n+1} : x_n > 0\}$, $R_0^n = \{x \in R^{n+1} : x_n = 0\}$,

$R_+^{n+1} = \{x \in R^{n+1} : x_0 > 0\}$, $Z^{n+1} = \mathbb{C}^1 \times R^{n-1}$,

$Z_+^{n+1} = \{\zeta' \in Z^{n+1} : \eta_0 \gtrless 0\}$;

$\alpha' = (\alpha_0, \alpha_1,\dots,\alpha_{n-1})$, $\alpha = (\alpha_0, \alpha_1,\dots,\alpha_{n-1},\alpha_n)$,

α_i - nonnegative integer, $|\alpha'| = \alpha_0 + \alpha_1 + \dots + \alpha_{n-1}$,

similarly $|\alpha|$;

$D_k = \frac{1}{i}\frac{\partial}{\partial x_k}$, $D^{\alpha'} = D_0^{\alpha_0} D_1^{\alpha_1} \dots D_{n-1}^{\alpha_{n-1}}$, similarly D^α.

1. The basic inequality

The main results of this paper are derived from a basic inequality for the proof of which this section is devoted.

We shall permanently use some norms and pseudodifferential operators depending on the parameter η_0. We shall define them.

In the space $\mathcal{H}_{(s)}(R^n)$ we introduce the norm

$$(1.1)\qquad \|u\|^2_{(s),\eta} = (2\pi)^{-n}\int |\zeta'|^{2s}\,|\hat{u}(\xi')|^2\,d\xi',\ \eta \neq 0.$$

Similarly in $\mathcal{H}_{(\mathfrak{z},s)}(R^{n+1})$

$$(1.2)\qquad \|u\|^2_{(\mathfrak{z},s),\eta} = (2\pi)^{-n-1}\int |\zeta|^{2\mathfrak{z}}\,|\zeta'|^{2s}\,|\hat{u}(\xi)|^2\,d\xi,\ \eta \neq 0.$$

The space $\mathcal{H}_{(\mathfrak{z},s)}(E^{n+1}_+)$ is equipped with the quotient norm

$$(1.3)\qquad \|u\|_{(\mathfrak{z},s),\eta} = \inf_U \|U\|_{(\mathfrak{z},s),\eta},$$

where inf is running over all extensions $U \in \mathcal{H}_{(\mathfrak{z},s)}(R^{n+1})$ of the distribution $u \in \mathcal{H}_{(\mathfrak{z},s)}(E^{n+1}_+)$.

In the sequel by $\|u\|_{(\mathfrak{z},s),\eta}$ we shall denote the norm in the last space, except the section 2 where it has also the meaning of the norm in $\overset{\circ}{\mathcal{H}}{}^+_{(\mathfrak{z},s)}(R^{n+1})$. It will be clear from the context in which sense it is used.

Finally we introduce the abbreviation

$$(1.4)\qquad |||u|||^2_{(k,s),\eta} = |\eta|\,\|u\|^2_{(k,s),\eta} + \sum_{l=0}^{k} \|D_n^l u(\cdot,0)\|^2_{(k-l+s),\eta}$$

where k is a nonnegative integer, $u \in \mathcal{H}_{(k+1,s-1)}(E^{n+1}_+)$ and

$D_n^l u(\cdot,0) \in \mathcal{H}_{(k-l+s)}(R^n)$.

For the definitions and properties of $\mathcal{H}_{(\delta,s)}(R^{n+1})$ and $\mathring{\mathcal{H}}^{+}_{(\delta,s)}(R^{n+1})$ the reader is referred to the book of L. Hörmander [4] .

We shall use the pseudodifferential operators $p(x',D'+i\eta')$ defined, as usually, by the equality

$$p(x',D'+i\eta')u(x') = (2\pi)^{-n}\int e^{i\langle x',\xi'\rangle} p(x',\zeta')\hat{u}(\xi')d\xi' ,$$

$$u \in C_0^\infty(R^n) , \quad \eta \neq 0 ,$$

where $p(x',\zeta') \in C^\infty\left(R^n \times (\overline{Z_-^{n+1}} \setminus \{0\})\right)$ is homogeneous of some degree r with respect to ζ' and constant outside a compact set with respect to x'. The variable η is a parameter in this definition. We shall use also the operators depending on additional variable x_n.

In the sequel by a pseudodifferential operator we shall understand an operator of the above class. We shall need the following properties of this class of operators.

(i) Let $p(x', D'+i\eta')$ and $q(x', D'+i\eta')$ be pseudodifferential operators of orders r and s respectively, and R_N is defined by

$$R_N = p(x', D'+i\eta')q(x', D'+i\eta') - \sum_{|\alpha|\leq N} \frac{1}{\alpha!} r_\alpha(x',D'+i\eta'),$$

where $r_\alpha(x', D'+i\eta')$ is the pseudodifferential operator with the symbol $r_\alpha(x', \zeta') = \partial^\alpha_{\xi'} p(x', \zeta') D^\alpha_{x'} q(x', \zeta')$.

For the operator R_N we have

$$\|R_N u\|_{(t),\eta} \leq K_N \|u\|_{(t+r+s-N-1),\eta}, \quad u \in C_0^\infty(R^n),$$

$|\eta| \geq 1$, s, N an arbitrary respectively real number and nonnegative integer.

(ii) Let $p(x', D' + i\eta')$ be a pseudodifferential operator of order r and $p^*(x', D' + i\eta')$ the adjoint operator, e.g. defined by the equality

$$\int p(x', D' + i\eta')u(x')\overline{v(x')}dx' =$$

$$= \int u(x')\, \overline{p^*(x', D' + i\eta')v(x')}dx', \quad u, v \in C_0^\infty(R^n).$$

Define the operator S_N by

$$S_N = p^*(x', D' + i\eta') - \sum_{|\alpha|=N} \frac{1}{\alpha!} p_\alpha(x', D' + i\eta'),$$

where $p_\alpha(x', D' + i\eta')$ is the pseudodifferential operator with the symbol $p_\alpha(x', \zeta') = D^\alpha_{x'} \partial^\alpha_{\xi'} \overline{p(x', \zeta')}$. We have

$$\|S_N u\|_{(s),\eta} \leq K_N \|u\|_{(s+r-N-1),\eta}, \quad u \in C_0^\infty(R^n),$$

$|\eta| \geq 1$, s, N an arbitrary respectively real number and nonnegative integer.

(iii) Let $P(x', D'+i\eta') = (p_{j,k}(x', D'+i\eta'))$,

$j,k = 1, \dots, m$, be a matrix of pseudodifferential operators all of order r, such that the matrix $P(x', \zeta') = (p_{j,k}(x', \zeta'))$ is hermitian. Let us suppose that

$$(P(x', \zeta')z, z) \geq \gamma_0 |\zeta'|^r |z|^2 \quad \text{for } z \in \mathbb{C}^m,\ \zeta' \in C_0' \cap \overline{Z_-^{n+1}},$$

where C_0' is an open cone in Z^{n+1}, and let $\psi \in C(Z^{n+1} \setminus \{0\})$ be a function homogeneous of degree 0 with $(\operatorname{supp} \psi) \setminus \{0\} \subset C_0'$. Then for each $\varepsilon > 0$

$$\operatorname{Re}\left(P(x', D'+i\eta')\psi(D'+i\eta')u, \psi(D'+i\eta')u\right) \geq$$

$$\geq (\gamma_0 - \varepsilon)\|\psi(D'+i\eta')u\|^2_{(\frac{r}{2}),\eta}$$

for all $u \in (C_0^\infty(R^n))^m$ and $|\eta|$ sufficiently large.

(iv) Let $p(x', D'+i\eta')$ be a pseudodifferential operator of order r and $\sup_{x', |\zeta'|=1} |p(x', \zeta')| = \gamma_0$, then for each ε

$$\|p(x', D'+i\eta')u\|_{(s),\eta} \leq (\gamma_0 + \varepsilon)\|u\|_{(r+s),\eta},$$

$u \in C_0^\infty(R^n)$,

for $|\eta|$ sufficiently large.

The proofs of these properties are similar to those given by Kohn and Nirenberg in the paper [7], where the reader may find basic facts on pseudodifferential operators.

Let $P(x,D)$ be a differential operator of order m defined on E_+^{n+1}, strongly hyperbolic with respect to x_0, which has smooth coefficients constant outside a compact subset of E_+^{n+1}. It is assumed that the hyperplane R_0^n is not characteristic for the operator $P(x,D)$, e.g.

$$P^0(x',0,\dots,0,1) = a_n(x') \neq 0 \quad \text{for } x' \in R_0^n .$$

$P^0(x,D)$ denotes the main part of the operator $P(x,D)$, similarly for another operator.

Since $a_n(x') \neq 0$ the polynomial $P^0(x',\zeta',\zeta_n)$ in ζ_n is exactly of degree m and thus it has m roots which change continuously with x' and ζ'.

If $\eta_0 < 0$ the equation $P^0(x',\zeta',\zeta_n) = 0$ has not real roots and by the continuous dependence on x', ζ' it follows that the number of roots with positive imaginary part is constant for all x', ζ'. Similarly for the number of roots with negative imaginary part. We have therefore

$$P^0(x',\zeta',\zeta_n) = a_n(x')\, Q^+(x',\zeta',\zeta_n) Q^-(x',\zeta',\zeta_n) ,$$

where $Q^+(x',\zeta',\zeta_n)$ and $Q^-(x',\zeta',\zeta_n)$ are polynomials with respect to the variable ζ_n with roots in upper and lower halfplane respectively. The coefficients of these polynomials are smooth functions of x', ζ', $\eta_0 < 0$, analytic with respect to ζ', they have continuous extensions for $\eta_0 \leq 0$. It follows from the Lemma 1 of this section.

Let the polynomial Q^+ be of degree $\varkappa$, $1 \leq \varkappa \leq m-1$. There are defined $\varkappa$ boundary operators on the hyperplane R_0^n

$$Q_j(x', D + i\eta) = \sum_{l=0}^{m_j} q_{j,l}(x', D' + i\eta')D_n^{m_j - l}, \quad j = 1,2,\ldots,\varkappa,$$

where $q_{j,l}(x', D' + i\eta')$ is a pseudodifferential operator of order l and $m_j \leq m - 1$.

Theorem 1. Let us suppose:

(A) for $x' \in R_0^n$, $\xi' \neq 0$ the real roots of the polynomial $P^0(x', \xi', \zeta_n)$ in ζ_n are at most of double multiplicity;

(B) the polynomials $Q_j(x', \zeta', \zeta_n)$ are linearly independent mod $Q^+(x', \zeta', \zeta_n)$ for $x' \in R_0^n$, $\zeta' \neq 0$, $\eta_0 \leq 0$.

Then there is a constant K such that

$$(1.5) \quad |\eta| \sum_{|\alpha|=m-1} \int_{E_+^{n+1}} |(D + i\eta)^\alpha u(x)|^2 dx + \sum_{|\alpha|=m-1} \int_{R_0^n} |(D + i\eta)u(x')|^2 dx' \leq$$

$$\leq K\Big(\int_{E_+^{n+1}} |P(x,D+i\eta)u(x)|^2 dx + \sum_{j=1}^{\varkappa} \sum_{|\alpha'|=m-1-m_j} \int_{R_0^n} |(D'+i\eta')^{\alpha'} Q_j(x',D+i\eta)u(x')|^2 dx' \Big)$$

for all $u \in C^{\infty}_{(o)}(E^{n+1}_{+})$ and $|\gamma|$ sufficiently large, $\gamma_0<0$. The constant K in the above inequality is independent of u and γ.

Proof. It is sufficient to prove the above inequality for the operator P^o instead of P (see [4], Chapter 8, Lemma 8.3.2).

We shall write P^o_{γ}, $Q_{j,\gamma}$ instead of $P^o(x, D + i\gamma)$ and $Q_j(x', D + i\gamma)$. Since $Q_j(x', \zeta, \zeta_n)$ may be replaced by $Q_j(x', \zeta', \zeta_n)|\zeta'|^{m-1-m_j}$ we can assume that $Q_{j,\gamma}$ is of order m-1, e.g.

$$Q_j(x',D + i\gamma) = \sum_{l=0}^{m-1} q_{j,l}(x', D' + i\gamma')D_n^{m-1-l},$$

where $q_{j,l}(x', D' + i\gamma')$ is a pseudodifferential operator of order l. Putting the meanings of the norms (1.1), (1.2) and (1.4) the inequality (1.5) is equivalent to

$$|||u|||^2_{(m-1,o),\gamma} \leq K\left(\|P^o_{\gamma}u\|^2_{(o,o),\gamma} + \sum_{j=1}^{\varkappa} \|Q_{j,\gamma}u(\cdot,0)\|^2_{(o),\gamma}\right).$$

Using a partition of unity we see that it is enough to prove this inequality locally, e.g. for each $x^o \in \overline{E^{n+1}_{+}}$ there exists a neighbourhood Ω_{x^o} of x^o such that the inequality is valid for all $u \in C^{\infty}_{o}(\Omega_{x^o})$. When $x^o \in E^{n+1}_{+}$, then the existence of such neighbourhood follows from the results

contained in the book [4] (Chapter 9, Theorem 9.2.1).
Thus it remains to prove it for $x^0 \in R^n_0$. We can assume that $x^0 = 0$. Let $\Omega_r = \{x \in R^{n+1} : |x| < r\}$, $\chi \in C_0^\infty(\Omega_2)$, $0 \leq \chi(x) \leq 1$ and $\chi(x) = 1$ for $x \in \Omega_1$, and let $\chi_\delta(x) = \chi(\delta^{-1}x)$, $\delta > 0$. Let $P^0_{\eta,\delta}$ and $Q_{j,\eta,\delta}$ are defined by

(1.6) $$P^0_{\eta,\delta}(x,D) = P^0(\chi_\delta(x)x, D + i\eta),$$

(1.7) $$Q_{j,\eta,\delta}(x',D) = Q_j(\chi_\delta(x')x', D + i\eta), \quad j = 1,\dots,\varkappa.$$

These operators coincide with P^0_η and $Q_{j,\eta}$ respectively in the set Ω_δ and their symbols in $\overline{E^{n+1}_+}$ are determined by the symbols of P^0_η and $Q_{j,\eta}$ in $\Omega_{2\delta}$.
It is enough to prove that for some $\delta > 0$ there exists a constant K such that the inequality

(1.8) $$|||u|||^2_{(m-1,0),\eta} \leq K(\|P^0_{\eta,\delta}u\|^2_{(0,0),\eta} + \sum_{j=1}^{\varkappa}\|Q_{j,\eta,\delta}u(\cdot,0)\|^2_{(0),\eta})$$

holds for all $u \in C^\infty_{(0)}(\overline{E^{n+1}_+})$ and $|\eta|$ sufficiently large.
To make the notations easier in further we abandon the indices η and δ in operators, and δ is always understand as "sufficiently small".
We can assume also that the polynomial $P^0(x',\zeta',\zeta_n)$ in ζ_n is normal.

The proof is divided in a few steps.

A.) Let ξ_n^0 be a real double root of the polynomial $P^0(0, \xi^{0\prime}, \zeta_n)$ in ζ_n for some $\xi^{0\prime} \neq 0$. The set of all points (x, ξ), where ξ_n is a double root, coincides with the set of the solutions of the two equations:

$$P^0(x, \xi) = 0 \quad , \quad \frac{\partial P^0(x,\xi)}{\partial \xi_n} = 0 \, .$$

Because

$$\begin{vmatrix} \dfrac{\partial P^0(0,\xi^0)}{\partial \xi_0} & \dfrac{\partial P^0(0,\xi^0)}{\partial \xi_n} \\[2ex] \dfrac{\partial^2 P^0(0,\xi^0)}{\partial \xi_0 \partial \xi_n} & \dfrac{\partial^2 P(0,\xi^0)}{\partial \xi_n^2} \end{vmatrix} = \frac{\partial P^0(0,\xi^0)}{\partial \xi_0} \cdot \frac{\partial^2 P^0(0,\xi^0)}{\partial \xi_n^2} \neq 0,$$

the point (x,ξ), ξ from some neighbourhood of ξ^0, is a solution of these equations if and only if $\xi_0 = \chi_0(x, \xi'')$, $\xi_n = \chi_n(x, \xi'')$, where χ_0 and χ_n are real smooth functions defined in $\overline{E_+^{n+1}} \times \omega_\varepsilon$, $\omega_\varepsilon = \{ \xi'' \in R^{n-1} : |\xi'' - \xi^{0\prime\prime}| < \varepsilon \}$.

Now we are going to investigate the structure of the set of all solutions of the equation $P^0(x, \zeta_0, \xi'', \zeta_n) = 0$ with respect to the variables ζ_0, ζ_n, in a complex neighbourhood of the point $(\chi_0(x,\xi''), \chi_n(x,\xi''))$. By the known theorem in the theory of analytic functions the solutions are of the form $(\zeta_0, z_n(x, \xi'', \sqrt{\zeta_0 - \chi_0}))$, where $z_n(x, \xi'', w)$ is an analytic function in w defined in some neighbourhood of 0, $z_n(x, \xi'', 0) =$

$= \chi_n(x, \xi'')$. Thus $P^o(x, \zeta_o, \xi'', z_n(x, \xi'', \sqrt{\zeta_o - \chi_o})) = 0$ and $P^o(x, w^2 + \chi_o, \xi'', z_n(x, \xi'', \pm w)) = 0$. Differentiating the last equation twicely in w we get for $w = 0$:

$$\left(\frac{\partial z_n(x, \xi'', 0)}{\partial w}\right)^2 = -2 \frac{\dfrac{\partial P^o(x, \chi_o, \xi'', \chi_n)}{\partial \xi_o}}{\dfrac{\partial P^o(x, \chi_o, \xi'', \chi_n)}{\partial \xi_n}} \neq 0 .$$

Hence it is clear that $z_n(x, \xi'', w)$ may be chosen so that either $a_1(x, \xi'')$, defined by

$$a_1(x, \xi'') = \frac{\partial z_n(x, \xi'', 0)}{\partial w} ,$$

or $-ia_1(x, \xi'')$ is positive. There is only one such function z_n. Let us write

$$z_n(x, \xi'', \sqrt{\zeta_o - \chi_o}) = \chi_n(x, \xi'') + \varphi(x, \xi'', a_1\sqrt{\zeta_o - \chi_o}).$$

The function $\varphi(x, \xi'', w)$ is analytic in w, $\varphi(x, \xi'', 0) = 0$, $\frac{\partial \varphi(x, \xi'', 0)}{\partial w} = 1$. It is important for us that $\varphi(x, \xi'', w)$ is a real valued if w is real. To prove it consider the case $a_1 > 0$. Let $\sqrt{\zeta_o - \chi_o}$ be the branch of the square root function which is positive on the halfline $\xi_o - \chi_o > 0$ and is discontinuous on the halfline $\xi_o - \chi_o \leq 0$. Because the polynomial $P^0(x, \zeta_0, \xi'', \zeta_n)$ in ζ_n has no real roots for $\Im \zeta_0 \neq 0$ it is $\Im z_n(x, \xi'', \sqrt{\zeta_0 - \chi_0}) = \Im \varphi(x, \xi'', a_1\sqrt{\zeta_0 - \chi_0}) \neq 0$ when $\Im \zeta_0 \neq 0$.

Since $\dfrac{\partial \varphi(x, \xi'', 0)}{\partial w} = 1$ and by the definition of $\sqrt{\zeta_0 - \chi_0}$

$\Im \varphi(x, \xi'', a_1 \sqrt{\zeta_0 - \chi_0}) > 0$ for $\Im \zeta_0 > 0$ and

$\Im \varphi(x, \xi'', a_1 \sqrt{\zeta_0 - \chi_0}) < 0$ for $\Im \zeta_0 < 0$, and so

$\Im \varphi(x, \xi'', a_1 \sqrt{\xi_0 - \chi_0}) = 0$ for $\xi_0 - \chi_0 > 0$. Thus the function $\varphi(x, \xi'', w)$ is real valued for $w > 0$ and since it is analytic it is real valued for all real w. If $-ia_1 > 0$ then choosing the branch of square root function which has values with positive imaginary part for $\xi_0 - \chi_0 < 0$ and which is discontinuous on the halfline $\xi_0 - \chi_0 \geqslant 0$ we can repeat the above argument.

Now we shall prove that $\varphi(x, \xi'', w)$ is a smooth function in the set $\overline{E_+^{n+1}} \times \omega_\varepsilon \times \{w \in \mathbb{C}^1 : |w| < \varepsilon\}$ for small $\varepsilon > 0$, analytic with respect to w.

Because $\dfrac{\partial P^0(x, \chi_0, \xi'', \chi_n)}{\partial \zeta_0} \neq 0$ there exist $\varepsilon_1 > 0$, $\varepsilon_2 > 0$ and a smooth function $z_0(x, \xi'', \zeta_n)$ defined on the set $\{(x, \xi'', \zeta_n) : x \in \overline{E_+^{n+1}}, \xi'' \in \omega_\varepsilon, |\zeta_n - \chi_n(x, \xi'')| < \varepsilon_1\}$, analytic in ζ_n, such that $|z_0(x, \xi'', \zeta_n) - \chi_0(x, \xi'')| < \varepsilon_2$, $P^0(x, z_0(x, \xi'', \zeta_n), \xi'', \zeta_n) = 0$ and if $|\zeta_0 - \chi_0(x, \xi'')| < \varepsilon_2$,

$|\zeta_n - \chi_n(x, \xi'')| < \varepsilon_1$, $P^o(x, \zeta_o, \xi'', \zeta_n) = 0$ then

$\zeta_o = z_o(x, \xi'', \zeta_n)$.

We have $\frac{\partial P^o}{\partial \zeta_o} \frac{\partial z_o}{\partial \zeta_n} + \frac{\partial P^o}{\partial \zeta_n} = 0$ and

$$\frac{\partial^2 P^o}{\partial \zeta_o^2}\left(\frac{\partial z_o}{\partial \zeta_n}\right)^2 + 2\frac{\partial^2 P^o}{\partial \zeta_o \partial \zeta_n}\frac{\partial z_o}{\partial \zeta_n} + \frac{\partial P^o}{\partial \zeta_o}\frac{\partial^2 z_o}{\partial \zeta_n^2} + \frac{\partial^2 P^o}{\partial \zeta_n^2} = 0 .$$

This gives $\frac{\partial z_o(x, \xi'', \chi_n)}{\partial \zeta_n} = 0$

and $$\frac{\partial^2 z_o(x, \xi'', \chi_n)}{\partial \zeta_n^2} = - \frac{\dfrac{\partial^2 P^o(x, \chi_o, \xi'', \chi_n)}{\partial \zeta_n^2}}{\dfrac{\partial P^o(x, \chi_o, \xi'', \chi_n)}{\partial \zeta_o}} \neq 0 .$$

Thus χ_n is a double root of the equation $z_o(x, \xi'', \zeta_n) - \chi_o = 0$. By a known theorem for analytic functions there exists $\varepsilon_2' > 0$ such that for each ζ_o with $0 < |\zeta_o - \chi_o| < \varepsilon_2'$ the equation $z_o(x, \xi'', \zeta_n) = \zeta_o$ has exactly two different solutions in the ring $0 < |\zeta_n - \chi_n| < \varepsilon_1$. Thus the inverse of z_o is an analytic double-valued function in the ring $0 < |\zeta_o - \chi_o| < \varepsilon_2'$, with algebraic critical point χ_o. This function coincides with the function $z_n(x, \xi'', \sqrt{\zeta_o - \chi_o})$ in some neighbourhood of χ_o, thus we can assume that

$z_n(x,\xi'',w)$ is defined on $\overline{E^{n+1}_+}\times\omega_\varepsilon\times\{w\in\mathbb{C}^1 : |w|<\sqrt{\varepsilon_2'}\}$.

Now let $(x^1,\xi^{1''})\in\overline{E^{n+1}_+}\times\omega_\varepsilon$, $0<|w_1|<\sqrt{\varepsilon_2'}$. There exists a number ζ_0^1 in the ring $0<|\zeta_0^1-\chi_0|<\varepsilon_2'$ such that $w_1=\sqrt{\zeta_0^1-\chi_0}$. Let $\zeta_n^1=z_n(x^1,\xi^{1''},w_1)$, then $0<|\zeta_n^1-\chi_n^1|<\varepsilon_1$, $P^0(x^1,\zeta_0^1,\xi^{1''},\zeta_n^1)=0$ and $\zeta_0^1=z_0(x^1,\xi^{1''},\zeta_n^1)$. If the number ε_1 is chosen sufficiently small so that

$$\frac{\partial z_0(x,\xi'',\zeta_n)}{\partial\zeta_n}\neq 0 \text{ in the ring } 0<|\zeta_n-\chi_n|<\varepsilon_1 ,$$

then $\dfrac{\partial z_0(x^1,\xi^{1''},\zeta_n^1)}{\partial\zeta_n}\neq 0$. Now again applying the implicit function theorem we have that the equation $z_0(x,\xi'',\zeta_n)-\zeta_0=0$ in sufficiently small neighbourhood of $(x^1,\zeta_0^1,\xi^{1''},\zeta_n^1)$ determines ζ_n as a smooth function of x,ξ'',ζ_0 in a neighbourhood of the point $(x^1,\xi^{1''},\zeta_0^1)$. This function coincides with $z_n(x,\xi'',\sqrt{\zeta_0-\chi_0})$. Because $\sqrt{\zeta_0-\chi_0}$ was chosen so that $\sqrt{\zeta_0-\chi_0}=w$ for $\zeta_0=\chi_0+w^2$ in some neighbourhood of w_1, we get that $z_n(x,\xi'',w)$ is a smooth function in a neighbourhood of the point $(x^1,\xi^{1''},w_1)$. Since the last point is quite arbitrary in the set $\overline{E^{n+1}_+}\times\omega_\varepsilon\times\{w\in\mathbb{C}^1 : 0<|w|<\sqrt{\varepsilon_2'}\}$, $z_n(x,\xi'',w)$ is a smooth function in this whole set. Since $z_n(x,\xi'',w)$ is analytic in w, the preceding implies that it is smooth in the set $\overline{E^{n+1}_+}\times\omega_\varepsilon\times\{w\in\mathbb{C}^1 : |w|<\sqrt{\varepsilon_2'}$. By the homogeneity of P^o the functions χ_o, χ_n, z_n are homogeneous of degree 1 with respect to ζ', so they are defined in the corresponding open cones.

Let us denote

$$C''_\varepsilon = \left\{ \xi'' \in R^{n-1} : \left| \xi'' - |\xi''|\,\xi^{o\prime\prime} \right| < \varepsilon |\xi''| \right\},$$

$$C'_{\varepsilon_o,\varepsilon}(x) = \left\{ \zeta' \in Z^{n+1} : \xi'' \in C''_\varepsilon,\ \left| \zeta_o - \chi_o(x,\xi'') \right| < \varepsilon_o |\xi''| \right\},$$

$$C_{\varepsilon_o,\varepsilon,\varepsilon_n}(x) = \left\{ \zeta \in Z^{n+1} \times \mathbb{C}^1 : \zeta' \in C'_{\varepsilon_o,\varepsilon}(x),\ \left| \zeta_n - \chi_n(x,\xi'') \right| < \varepsilon_n |\xi''| \right\},$$

and let $\sqrt{\zeta_o - \chi_o}$ be the branch of the square root function which has values with the positive imaginary part for $\xi_o - \chi_o < 0$ and which is splitted along the ray $\xi_o - \chi_o \geqslant 0$.

All the above considerations may be summarized in the following proposition.

There exist ε_o, ε and ε_n such that the set of all solutions of the equation $P^o(x,\zeta',\zeta_n) = 0$ from the set $C_{\varepsilon_o,\varepsilon,\varepsilon_n}(x) \cap \overline{Z^{n+1}_-}$ is given by two equations :

$$(1.9)\quad \zeta_n = z_n^{\pm}(x,\xi'',\zeta_o) = \chi_n(x,\xi'') + \varphi\left(x,\xi'', \pm a_1(x,\xi'')\sqrt{\zeta_o - \chi_o(x,\xi'')}\right)$$

The functions χ_o, χ_n and a_1 are smooth in $\overline{E^{n+1}_+} \times C''_\varepsilon$ and $\chi_o(x,\tau\xi'') = \tau\chi_o(x,\xi'')$, $\chi_n(x,\tau\xi'') = \tau\chi_n(x,\xi'')$, $a_1(x,\tau\xi'') = \sqrt{\tau}\, a_1(x,\xi'')$, $\tau > 0$. Either the function a_1 is positive, or $-ia_1$ is positive.

The function $\varphi(x, \xi'', w)$ is smooth, real valued for w real and it is analytic in w in the set

$$\left\{(x, \xi'', w) : (x, \xi'') \in \overline{E_+^{n+1}} \times C''_\varepsilon , |w| < |a_1(x, \xi'')| \sqrt{\varepsilon_0 |\xi''|}\right\}$$

moreover $\varphi(x, \xi'', 0) = 0$, $\dfrac{\partial(x, \xi'', 0)}{\partial w} = 1$ and

$$\varphi(x, \tau\xi'', \tau w) = \tau\varphi(x, \xi'', w) , \quad \tau > 0.$$

Hence

$$\operatorname{Im} z_n^+(x, \xi'', \zeta_0) > 0 \text{ and } \operatorname{Im} z_n^-(x, \xi'', \zeta_n) < 0 .$$

B.) Lemma 1. Let $p^o(z) = z^m + w_1^o z^{m-1} + \ldots + w_m^o$, $w^o = (w_1^o , \ldots , w_m^o) \in \mathbb{C}^m$, and let γ be a Jordan curve disjoint with the roots of the polynomial $p^o(z)$, which bounds k roots (taking into account their multiplicity). There exists $\varepsilon > 0$ such that for $w \in \prod(w^o, \varepsilon) = \{w \in \mathbb{C}^m : |w_i - w_i^o| < \varepsilon, i=1,\ldots,m\}$ all roots of the polynomial $p(z) = z^m + w_1 z^{m-1} + \ldots + w_m$ are disjoint with γ , γ bounds exactly k roots $\lambda_1, \ldots, \lambda_k$ and the coefficients of the polynomial $q(z) = (z-\lambda_1)\ldots(z-\lambda_k)$ are analytic functions for $w \in \prod(w^o, \varepsilon)$.

Proof of the lemma. Only the last statement requires arguments. We can write $q(z) = z^k - \sigma_1(\lambda_1, \ldots, \lambda_k) z^{k-1} + \sigma_2(\lambda_1, \ldots, \lambda_k) z^{k-2} - \ldots + (-1)^k \sigma_k(\lambda_1, \ldots, \lambda_k)$, where

$$\sigma_j(\lambda_1,\ldots,\lambda_k) = \sum_{r_1<r_2<\ldots<r_j} \lambda_{r_1}\lambda_{r_2}\ldots\lambda_{r_j} ,$$

$j = 1,2,\ldots,k$. So $\sigma_1(\lambda_1,\ldots,\lambda_k) = \lambda_1 + \ldots + \lambda_k =$

$= \frac{1}{2\pi i}\int_\gamma \frac{zp'(z)}{p(z)}dz$, and this function is analytic in $\Pi(w^0,\varepsilon)$. For σ_2 we have:

$$\sigma_2(\lambda_1,\ldots,\lambda_k) = \sum_{i<j}\lambda_i\lambda_j = \frac{1}{2}\sum_{i\neq j}\lambda_i\lambda_j = \frac{1}{2}(\sum_{i,j}\lambda_i\lambda_j - \sum_i\lambda_i^2) =$$

$$= \frac{1}{2}((\sum_i\lambda_i)^2 - \sum_i\lambda_i^2) = \frac{1}{2}((\sigma_1(\lambda_1,\ldots,\lambda_k))^2 - \frac{1}{2\pi i}\int_\gamma \frac{z^2p'(z)}{p(z)}dz)$$

hence analyticity of σ_2. The proof for the other coefficients is similar.

Let $\xi^{0\prime}\in R^n$, $|\xi^{0\prime}| = 1$ and let ξ_0^0 be arbitrary. Consider the polynomial $P^0(0,\xi^{0\prime},\zeta_n)$ in ζ_n. It has p roots in the upper halfplane, q in the lower halfplane, r real simple roots which are simultaneously the roots of the polynomial $Q^+(0,\xi^{0\prime},\zeta_n)$, s real simple roots which are simultaneously the roots of $Q^-(0,\xi^{0\prime},\zeta_n)$, and finally t real double roots. It follows from Lemma 1 that in the set $\overline{E^{n+1}_+}\times\{\zeta_0\in\mathbb{C}^1 : |\zeta_0-\xi_0^0|<\varepsilon\}\times\omega_\varepsilon$ we have the factorisation:

$$(1.10)\quad P^0(x,\zeta',\zeta_n) = P^+(x,\zeta',\zeta_n)P^-(x,\zeta',\zeta_n)\prod_{j=1}^{r} M_j^+(x,\zeta',\zeta_n)\cdot$$

$$\cdot\prod_{k=1}^{s} M_k^-(x,\zeta',\zeta_n)\prod_{l=1}^{t} N_l(x,\zeta',\zeta_n) .$$

The polynomials in this factorisation are normal and their coefficients are smooth functions, analytic in ζ'. Since the homogeneity of P^0 the equality (1.10) can be extended on the suitable cones, thus on the set $\overline{E_+^{n+1}} \times C_0'$, where

$$C_0' = \left\{ \zeta' \in Z^{n+1} : \left| \zeta_0 - \xi_0^0 \left| \xi'' \right| \right| < \varepsilon \left| \xi'' \right|, \ \xi'' \in C_\varepsilon'' \right\}.$$

The numbers δ and ε can be chosen so that the cone C_0' is contained in each cone $C'_{\varepsilon_0(l),\varepsilon(l)}(x)$, $x \in \overline{E_+^{n+1}}$, $l = 1,\dots,t$, constructed in the point A.) for the real double roots of $P^0(0, \xi^{0\prime}, \zeta_n)$.

The polynomials in the factorisation (1.10) have the following properties:

a) P^+ and P^- are the polynomials of degree p and q respectively, all roots of P^+ are in the upper half-plane, all roots of P^- are in the lower halfplane, thus

$$P^+(x, \zeta', \zeta_n) \neq 0 \quad \text{for} \quad \Im m\, \zeta_n \leq 0 ,$$

$$P^-(x, \zeta', \zeta_n) \neq 0 \quad \text{for} \quad \Im m\, \zeta_n \geq 0 ,$$

and their coefficients are homogeneous functions in ζ' of suitable degree.

b) M_j^+ and M_k^- are polynomials of degree 1 ,

$M_j^+(x,\zeta',\zeta_n) = \zeta_n - \lambda_j^+(x,\zeta')$, $\operatorname{Im}\lambda_j^+(x,\zeta') > 0$

for $\eta_0 < 0$,

$\lambda_j^+(x,\zeta')$ are homogeneous functions of degree 1 in ζ'

and $\lambda_j^+(x,\xi')$ are real valued ;

$M_k^-(x,\zeta',\zeta_n) = \zeta_n - \lambda_k^-(x,\zeta')$, $\operatorname{Im}\lambda_k^-(x,\zeta') < 0$ for $\eta_0 < 0$,

$\lambda_k^-(x,\zeta')$ are homogeneous functions of degree 1 in ζ'

and $\lambda_k^-(x,\xi')$ are real valued.

c) N_1 are polynomials of degree 2 ,

$$N_1(x,\zeta',\zeta_n) = \zeta_n^2 + p_1(x,\zeta')\zeta_n + q_1(x,\zeta') ,$$

$p_1(x,\zeta')$ and $q_1(x,\zeta')$ are homogeneous functions of degree 1 and 2 respectively in ζ' . We have also

$N_1(x,\zeta',\zeta_n) = (\zeta_n - z_{n,1}^+(x,\zeta'))(\zeta_n - z_{n,1}^-(x,\zeta'))$

for $\eta_0 \leq 0$,

where the functions $z_{n,1}^{\pm}$ are described in the conclusion of the point A.) . Their properties imply that $p_1(x,\xi')$, $q_1(x,\xi')$ are real valued functions.

The last statement in (c) requires some arguments. We have

$$p_1(x,\zeta') = -(z^+_{n,1}(x,\zeta') + z^-_{n,1}(x,\zeta')) =$$

$$= -2\chi_{n,1}(x,\xi'') - (\varphi_\iota(x,\xi'', \overline{a_{1,1}}\sqrt{\zeta_0 - \chi_{0,\iota}}) +$$

$$+ \varphi_\iota(x,\xi'', \overline{-a_{1,1}}\sqrt{\zeta_0 - \chi_{0,\iota}})),$$

$$q_1(x,\zeta') = z^+_{n,1}(x,\zeta')z^-_{n,1}(x,\zeta') = \chi^2_{n,1}(x,\xi'') +$$

$$+\chi_{n,1}(x,\xi'')(\varphi_\iota(x,\xi'',\overline{a_{1,1}}\sqrt{\zeta_0 - \chi_{0,\iota}}) +$$

$$+ \varphi_\iota(x,\xi'', \overline{-a_{1,1}}\sqrt{\zeta_0 - \chi_{0,\iota}})) +$$

$$+ \varphi_\iota(x,\xi'',\overline{a_{1,1}}\sqrt{\zeta_0 - \chi_{0,\iota}})\,\varphi_\iota(x,\xi'',\overline{-a_{1,1}}\sqrt{\zeta_0 - \chi_{0,\iota}}).$$

There are two possibilities: either $a_{1,1}$ is the real function, or $-a_{1,1}$ is the real function. Consider the first one. Then $\overline{-a_{1,1}}\sqrt{\xi_0 - \chi_{0,\iota}} = \overline{a_{1,1}\sqrt{\xi_0 - \chi_{0,\iota}}}$ for $\xi_0 - \chi_{0,\iota} < 0$ and since $\varphi_\iota(x,\xi'',\overline{w}) = \overline{\varphi_\iota(x,\xi'',w)}$ we have

$$p_1(x,\xi') = -2\chi_{n,1}(x,\xi'') - 2\mathrm{Re}\,\varphi_\iota(x,\xi'', \overline{a_{1,1}}\sqrt{\xi_0 - \chi_{0,\iota}}),$$

$$q_1(x,\xi') = -\chi^2_{n,1}(x,\xi'') - \chi_{n,1}(x,\xi'')p_1(x,\xi') +$$

$$+ \left| \varphi_l(x, \xi'', \overline{a_{1,1} \sqrt{\xi_0 - \chi_{0,l}}}) \right|^2 .$$

Thus $p_1(x,\xi')$, $q_1(x,\xi')$ are real valued for $\xi_0 - \chi_{0,l}(x,\xi'') < 0$, but the analyticity in ξ' implies that they are real valued for all ξ'.

If $-ia_{1,1}$ is a real function, then $-\overline{a_{1,1}} \sqrt{\xi_0 - \chi_{0,l}} =$

$= \overline{a_{1,1} \sqrt{\xi_0 - \chi_{0,l}}}$ for $\xi_0 - \chi_{0,l} > 0$ and further arguments go on without changes.

From the equality (1.10) and the properties (a), (b) and (c) it follows that

$$Q^+(x,\zeta',\zeta_n) = P^+(x,\zeta',\zeta_n) \prod_{j=1}^{r} M_j^+(x,\zeta',\zeta_n) \prod_{l=1}^{t} (\zeta_n - z_{n,l}^+(x,\zeta')),$$

$$Q^-(x,\zeta',\zeta_n) = P^-(x,\zeta',\zeta_n) \prod_{k=1}^{s} M_k^-(x,\zeta',\zeta_n) \prod_{l=1}^{t} (\zeta_n - z_{n,l}^-(x,\zeta'))$$

for $\eta_0 \leq 0$,

and $\varkappa = p + r + t$, $m - \varkappa = q + s + t$.

Introduce the denotations :

$$R^+ = \frac{P^0}{P^+}, \quad R^- = \frac{P^0}{P^-}, \quad R_j^+ = \frac{P^0}{M_j^+}, \quad j = 1,\dots,r ,$$

$$R_k^- = \frac{P^0}{M_k^-}, \quad k = 1,\dots,s , \quad R_l = \frac{P^0}{N_l}, \quad l = 1,\dots,t ,$$

and let us write $B_1, B_2, \dots, B_m$ for the polynomials

$R^+\zeta_n^{p-1}$, $R^+|\zeta'|\zeta_n^{p-2}$, ..., $R^+|\zeta'|^{p-1}$, R_1^+, $R_2^+, \dots, R_r^+$, $R_1|\zeta'|$,

$R_2|\zeta'|, \dots$, $R_t|\zeta'|$, $R_1\zeta_n$, $R_2\zeta_n, \dots, R_t\zeta_n$, $R^-\zeta_n^{q-1}$, $R^-|\zeta'|\zeta_n^{q-2}$, ...,

$R^-|\zeta'|^{q-1}$, R_1^-, $R_2^-, \dots$, R_s^- respectively.

C.) Now we shall prove that for sufficiently small δ and ε, and for each $(x,\zeta') \in \overline{E_+^{n+1}} \times C_o'$ the polynomials $\{B_j\}$, $j = 1,\dots,m$, form a basis in the m-dimensional vector space V_m of the all polynomials in ζ_n of degree $< m$.

Utilizing the continuity and the homogeneity of the coefficients of the polynomials B_j we can reduce the proof of the above statement for the general (x, ζ') to the proof for the point $(0, \xi^{o\prime})$. Thus let

$$(1.10a) \qquad \sum_{j=1}^{m} c_j B_j(0, \xi^{o\prime}, \zeta_n) = 0.$$

Substituting the roots of the polynomials $M_j^+(0,\xi^{o\prime},\zeta_n)$ and $M_k^-(0,\xi^{o\prime},\zeta_n)$ we get that the coefficients c_j at the polynomials $R_j^+(0,\xi^{o\prime},\zeta_n)$ and $R_k^-(0,\xi^{o\prime},\zeta_n)$ are equal to 0.

Next substitute ν_l^o - the double root of $N_1(0,\xi^{o\prime},\zeta_n)$.

We have the equation :

$$(1.11) \quad c_{j(1)}R_1(0,\xi^{o\prime},\nu_l^o)|\xi^{o\prime}| + c_{j(1)+t}R_1(0,\xi^{o\prime},\nu_l^o)\nu_l^o = 0,$$

$$R_1(0,\xi^{o\prime},\nu_l^o) \neq 0.$$

The differentiation of (1.10a) with respect to ζ_n and the substitution of ν_l^o gives the second equation:

$$(1.12)\quad c_{j(l)} \frac{\partial R_l(0,\xi^{0\prime},\nu_l^0)}{\partial \zeta_n}|\xi^{0\prime}| + c_{j(l)+t} \frac{\partial R_l(0,\xi^{0\prime},\nu_l^0)}{\partial \zeta_n}\nu_l^0 +$$

$$+ c_{j(l)+t} R_l(0,\xi^{0\prime},\nu_l^0) = 0 .$$

The determinant of the system of the equations (1.11) and (1.12) is equal to

$$\begin{vmatrix} R_l(0,\xi^{0\prime},\nu_l^0)|\xi^{0\prime}| , & R_l(0,\xi^{0\prime},\nu_l^0)\nu_l^0 \\ \dfrac{\partial R_l(0,\xi^{0\prime},\nu_l^0)}{\partial \zeta_n}|\xi^{0\prime}| , & \dfrac{\partial R_l(0,\xi^{0\prime},\nu_l^0)}{\partial \zeta_n}\nu_l^0 + R_l(0,\xi^{0\prime},\nu_l^0) \end{vmatrix} = R_l^2(0,\xi^{0\prime},\nu_l^0)|\xi^{0\prime}| \neq 0,$$

hence $c_{j(l)} = c_{j(l)+t} = 0$.

From this we obtain the equation

$$(1.13)\quad C^+(\zeta_n)R^+(0,\xi^{0\prime},\zeta_n) + C^-(\zeta_n)R^-(0,\xi^{0\prime},\zeta_n) = 0 ,$$

where $C^+(\zeta_n) = c_1\zeta_n^{p-1} + c_2|\zeta'|\zeta_n^{p-2} + \dots + c_p|\zeta'|^{p-1}$,

$$C^-(\zeta_n) = c_{\varkappa+t+1}\zeta_n^{q-1} + c_{\varkappa+t+2}|\zeta'|\zeta_n^{q-2} + \dots +$$

$$+ c_{\varkappa+t+q}|\zeta'|^{q-1} .$$

Dividing (1.13) by the common factor of R^+ and R^- we have

$$C^+(\zeta_n)P^-(0,\xi^{0\prime},\zeta_n) + C^-(\zeta_n)P^+(0,\xi^{0\prime},\zeta_n) = 0 .$$

We can write

$$C^{+}(\zeta_n) = -\frac{C^{-}(\zeta_n)P^{+}(0,\xi^{0\prime},\zeta_n)}{P^{-}(0,\xi^{0\prime},\zeta_n)} \qquad \text{for } \operatorname{Im}\zeta_n > 0 ,$$

so it follows that the polynomial $C^{+}(\zeta_n)$ has p roots in the upper halfplane. It implies that $C^{+}(\zeta_n)$ must be identically equal to 0, thus $c_1 = c_2 = \dots = c_p = 0$. Similarly we get $c_{\varkappa+t+1} = \dots = c_{\varkappa+t+q} = 0$ and therefore the polynomials $\left\{B_j(0,\xi^{0\prime},\zeta_n)\right\}$, $j = 1,\dots,m$, are linearly independent.

From the property C.) it follows that

$$(1.14) \qquad |\zeta'|^{m-1-j}\zeta_n^{j} = \sum_{k=1}^{m} a_{j,k}(x,\zeta')B_k(x,\zeta',\zeta_n) ,$$

$$j = 0,1,\dots,m-1 ,$$

and

$$(1.15) \qquad Q_j(x',\zeta',\zeta_n) = \sum_{k=1}^{m} b_{j,k}(x',\zeta')B_k(x',\zeta',\zeta_n) ,$$

$$j = 1,2,\dots,\varkappa .$$

Here $a_{j,k}(x,\zeta')$ and $b_{j,k}(x',\zeta')$ are smooth and homogeneous of degree 0 functions defined on $\overline{E_{+}^{n+1}} \times C_o'$ and $R_o^n \times C_o'$ respectively.

We shall now investigate consequences of the assumption (B).

Let us write the equality (1.15) in the following form:

$$(1.16)\qquad Q_j(x,\zeta',\zeta_n) = \sum_{k=1}^{p+r} b_{j,k}(x,\zeta')B_k(x,\zeta',\zeta_n) +$$
$$+ \sum_{k=p+r+1}^{\varkappa} (b_{j,k}(x,\zeta')|\zeta'| +$$
$$+ b_{j,k+t}(x,\zeta')z^+_{n,k-p-r}(x,\zeta')\, R_{k-p-r}(x,\zeta',\zeta_n) +$$
$$+ \sum_{k=\varkappa+1}^{\varkappa+t} b_{j,k}(x,\zeta')(\zeta_n - z^+_{n,k-\varkappa}(x,\zeta'))R_{k-\varkappa}(x,\zeta',\zeta_n) +$$
$$+ \sum_{k=\varkappa+t+1}^{m} b_{j,k}(x,\zeta')B_k(x,\zeta',\zeta_n)\ , \qquad j = 1,\dots,\varkappa\ ,$$

The polynomials $B_1,\dots,B_{p+r}, R_1,\dots,R_t, (\zeta_n - z^+_{n,1})R_1,\dots,$ $(\zeta_n - z^+_{n,t})R_t, B_{\varkappa+t+1},\dots,B_m$ form a basis in the space V_m. Because $(\zeta_n - z^+_{n,1})R_1,\dots,(\zeta_n - z^+_{n,t})R_t, B_{\varkappa+t+1},\dots,B_m$ are divisible by Q^+, it follows from the assumption (B) and (1.16) that the polynomials

$$\sum_{k=1}^{p+r} b_{j,k}B_k + \sum_{k=p+r+1}^{\varkappa} (b_{j,k}|\zeta'| + b_{j,k+t}z^+_{n,k-p-r})R_{k-p-r} =$$
$$= \sum_{k=1}^{p+r} b_{j,k}B_k + \sum_{k=p+r+1}^{\varkappa} (b_{j,k} + b_{j,k+t}\frac{z^+_{n,k-p-r}}{|\zeta'|})B_k \overset{def}{=}$$
$$\overset{def}{=} \sum_{k=1}^{\varkappa} b'_{j,k}B_k, \qquad j = 1,\dots,\varkappa\ ,$$

are linearly independent.

Thus $\det(b_{j,k}(x,\zeta'))_{j,k=1}^{\varkappa} \neq 0$. Let $C'(x,\zeta') = (c_{i,j}(x,\zeta'))_{i,j=1}^{\varkappa}$ be the inverse matrix of the matrix $(b'_{j,k}(x,\zeta'))_{j,k=1}^{\varkappa}$.

Then we have :

$$(1.17)\quad \sum_{j=1}^{\varkappa} c'_{i,j}(x,\zeta')Q_j(x,\zeta',\zeta_n) = B_i(x,\zeta',\zeta_n) +$$

$$+\sum_{k=\varkappa+1}^{\varkappa+t}\Big(\sum_{j=1}^{\varkappa} c'_{i,j}(x,\zeta')b_{j,k}(x,\zeta')\Big)(\zeta_n - z^+_{n,k-\varkappa}(x,\zeta'))R_{k-\varkappa}(x,\zeta',\zeta_n)+$$

$$+\sum_{k=\varkappa+t+1}^{m}\Big(\sum_{j=1}^{\varkappa} c'_{i,j}(x,\zeta')b_{j,k}(x,\zeta')\Big)B_k(x,\zeta',\zeta_n)$$

for $i = 1,\dots,\varkappa$.

The matrix $C'(x,\zeta')$ is not suit for our purposes because its elements are not smooth for $\eta_0 = 0$, they are only continuous here, and therefore in general they are not symbols of pseudodifferential operators. In the future we shall use only the matrix $C'(0,\xi^{0\prime})$.

D.) Let $\psi,\psi_1 \in C^\infty(Z^{n+1}\setminus\{0\})$ be functions homogeneous of degree 0 with $(\operatorname{supp}\psi)\setminus\{0\}\subset C'_0$, $(\operatorname{supp}\psi_1)\setminus\{0\}\subset C'_0$ and $\psi_1(\zeta') = 1$ for $\zeta'\in(\operatorname{supp}\psi)\setminus\{0\}$.

In the preceding points were considered the smooth and homogeneous functions defined on the set $\overline{E^{n+1}_+}\times C'_0$. Multiplying them by $\psi_1(\zeta')$ we obtain the symbols of the pseudodifferential operators. Such operators shall be applied to the functions $\psi(D' + i\eta')u$, shortly ψu, where $u \in C^{\infty}_{(o)}(E^{n+1}_{+})$. In this point we shall prove the following inequalities for the operators $P^{+}\psi_1(x, D + i\eta)$ and $P^{-}\psi_1(x, D + i\eta)$:

$$(1.18) \qquad |\eta|\,\|\psi u\|^2_{(p-1,o),\eta} \leq K\Big(\int_{E^{n+1}_{+}} |P^{+}\psi_1(x,D + i\eta)\,\psi u|^2\,dx + \sum_{j=0}^{p-1}\|D^j_n\psi u(\cdot,0)\|^2_{(p-1-j),\eta}\Big),$$

$$(1.19) \qquad |\eta|\,\|\psi u\|^2_{(q-1,0),\eta} + \sum_{j=0}^{q-1}\|D^j_n\psi u(\cdot,0)\|^2_{(q-1-j),\eta} \leq K\int_{E^{n+1}_{+}} |P^{-}\psi_1(x,D + i\eta)\psi u|^2\,dx$$

for all $u \in C^{\infty}_{(o)}(E^{n+1}_{+})$ and $|\eta|$ sufficiently large, K is a constant independent of u and η. Consider the operator $P^{+}(0, \zeta', D_n)$.

The roots of the polynomial $P^{+}(0, \zeta', \zeta_n)$ are in the upper halfplane, hence the vector space V_p of solutions of the equation $P^{+}(0, \zeta', D_n)v(x_n) = 0$, defined and bounded on $\overline{R^1_{+}}$, is p-dimensional and the mapping

$$T_o^{-1} : v \in V_p \longrightarrow (v(o), D_n v(0), \ldots, D_n^{p-1}\, v(0)) \in \mathbb{C}^p$$

is an isomorphism of the spaces V_p and $\mathbb{C}^p$. Consider the mapping

$$T_1 : f \in \mathring{\mathcal{H}}^{+}_{(0)}(R^1) \longrightarrow (2\pi)^{-1} \int \frac{e^{ix_n \xi_n} \hat{f}(\xi_n)}{P^{+}(0, \zeta', \xi_n)} d\xi_n \in$$

$$\in \mathring{\mathcal{H}}^{+}_{(p)}(R^1) .$$

The last inclusion above holds because the functions $\hat{f}(\xi_n)$ and $\frac{1}{P^{+}(0, \zeta', \xi_n)}$ can be prolonged on the lower halfplane to the analytic functions with the suitable estimates as it follows from the definition of $\mathring{\mathcal{H}}^{+}_{(0)}(R^1)$ and the property (a). T_1 is a continuous mapping and $P^{+}(0, \zeta', D_n)T_1 = I$, similarly $T_1 P^{+}(0, \zeta', D_n) = I$, hence T_1 and $P^{+}(0, \zeta', D_n)$ are the isomorphisms of the suitable spaces.

Define the mappings

$$T : (f, v_0, v_1, \ldots, v_{p-1}) \in \mathcal{H}_{(0)}(R^1_+) \oplus \mathbb{C}^p \longrightarrow T_0 v + T_1 f \in \mathcal{H}_{(p)}(R^1_+)$$

and

$$A^{+} : u \in \mathcal{H}_{(p)}(R^1_+) \longrightarrow (P^{+}(0, \zeta', D_n)u, u(0), D_n u(0), \ldots, D_n^{p-1} u(0)) \in$$

$$\in \mathcal{H}_{(0)}(R^1_+) \oplus \mathbb{C}^p .$$

These mappings are continuous , $A^{+}T = I$ and if $A^{+}u = 0$ then $u = 0$. Thus $TA^{+} = I$, T and A^{+} are isomorphisms and it implies that the inequality

$$\sum_{j=0}^{p}\int_0^{+\infty}|D_n^j u(x_n)|^2\,dx_n \leq K\Big(\int_0^{\infty}|P^+(0,\zeta',D_n)u(x_n)|^2\,dx_n + \tag{1.20}$$
$$+\sum_{j=0}^{p-1}|D_n^j u(0)|^2\Big)$$

holds for all $u \in C^{\infty}_{(o)}(R^1_+)$.

The best constant K in this inequality is the lower semi-continuous function of ζ', therefore it can be chosen common for all $\zeta' \in C'_o$, $|\zeta'| = 1$. Applying (1.20) for $\frac{\zeta'}{|\zeta'|}$, $\zeta' \in C'_o$, and for the function $\Psi(\zeta')\hat{u}(\xi', \frac{x_n}{|\zeta'|})$, and multiplying so obtained inquality by $|\zeta'|^{2p}$ and integrating with respect to ξ' we have

$$\|\Psi u\|^2_{(p,o),\eta} \leq K\Big(\int_{E^{n+1}_+}|P^+\,\Psi_1(0,D+i\eta)\Psi u|^2\,dx + \tag{1.21}$$
$$+\sum_{j=0}^{p-1}\|D_n^j\Psi u(\cdot,0)\|^2_{(p-j-\frac{1}{2}),\eta}\Big)$$

for $u \in C^{\infty}_{(o)}(E^{n+1}_+)$.

Let us write $P^+(x,\zeta,\zeta_n) = \zeta_n^{\,p} + a_1(x,\zeta')\zeta_n^{p-1} + \ldots + a_p(x,\zeta')$.

The definition (9) implies that

$$\sup_{x,|\zeta'|=1}|a_i(x,\zeta')\Psi_1(\zeta') - a_i(0,\zeta')\Psi_1(\zeta')| \longrightarrow 0 \text{ when } \delta \to 0^+ .$$

From the proposition (iv) it follows that

$$\int_{R^n}|a_i\Psi_1(x,D'+i\eta')v(x') - a_i\Psi_1(0,D'+i\eta')v(x')|^2dx' \leq O(\delta^2)\|v\|^2_{(i),\eta}$$

for $|\eta| \geq \tau(\delta)$ and $v \in \mathcal{H}_{(i)}(R^n)$. Therefore

$$\int_{E_+^{n+1}} |P^+ \psi_1(x,D+i\eta)\psi u - P^+\psi_1(0,D+i\eta)\psi u|^2 dx \leq O(\delta^2)\|\psi u\|^2_{(p,o),\eta}$$

for $|\eta| \geq \tau_1(\delta)$ and $u \in C^\infty_{(o)}(E_+^{n+1})$.

This implies for sufficiently small δ the inequality (1.21) for the operator $P^+\psi_1(x,D+i\eta)$ instead of the operator $P^+\psi_1(0, D+i\eta)$.

Denote by $\Lambda_{(s),\eta}$ the pseudodifferential operator with the symbol $|\xi'|^s$ and substitute in the above inequality the function $\Lambda_{(-\frac{1}{2}),\eta}u$ in the place of the function u.

By the proposition (i) we have

$$\int_{E_+^{n+1}} |P^+\psi_1(x,D+i\eta)\Lambda_{(-\frac{1}{2}),\eta}\psi u - \Lambda_{(-\frac{1}{2}),\eta} P^+\psi_1(x,D+i\eta)\psi u|^2 dx \leq$$

$$\leq K_1\|\psi u\|^2_{(p-1,\frac{1}{2}),\eta} ,$$

hence

$$(1.22) \quad \|\psi u\|^2_{(p,-\frac{1}{2}),\eta} \leq K\Big(\|(P^+\psi_1)\psi u\|^2_{(0,-\frac{1}{2}),\eta} +$$

$$+ \sum_{j=0}^{p-1} \|D_n^j \psi u(\cdot\,,0)\|^2_{(p-1-j),\eta} +$$

$$+ \|\psi u\|^2_{(p-1,-\frac{1}{2}),\eta}\Big) ,$$

$u \in C^{\infty}_{(o)}(E^{n+1}_{+})$, $|\eta|$ large .

Estimating $\|\psi u\|^2_{(p,-\frac{1}{2}),\eta} \geq \|\psi u\|^2_{(p-1,\frac{1}{2}),\eta} \geq |\eta| \|\psi u\|^2_{(p-1,0),\eta}$

we see that the last term on the right in (1.22) can be omitted for $|\eta|$ large. Because

$$\|(P^{+}\psi_1)\psi u\|^2_{(0,-\frac{1}{2}),\eta} \leq \|(P^{+}\psi_1)\psi u\|^2_{(o,o),\eta}$$

the inequality (1.18) is proved.

The proof of (1.19) is very similar and we shall only sketch it.

The equation $P^{-}(0,\zeta',D_n)v(x_n) = 0$ has not solutions bounded on R^1_+ and for $f \in \mathcal{H}_{(o)}(R^1_+)$ the formula

$$u(x_n) = (2\pi)^{-1} \int \frac{e^{ix_n\xi_n} f(\xi_n)}{P^{-}(0,\zeta',\xi_n)} d\xi_n$$

defines the unique solution from $\mathcal{H}_{(q)}(R^1_+)$ of the equation $P^{-}(0,\zeta',D_n)u = f$. Thus the operator $P^{-}(0,\zeta',D_n)$ is an isomorphism of the space $\mathcal{H}_{(q)}(R^1_+)$ with $\mathcal{H}_{(o)}(R^1_+)$ and the inequality

$$\sum_{j=0}^{q} \int_0^{+\infty} |D_n^j u(x_n)|^2 dx_n \leq K \int_0^{+\infty} |P^{-}(0,\zeta',D_n)u(x_n)|^2 dx_n$$

holds for $u \in C^{\infty}_{(o)}(R^1_+)$, hence also

$$\sum_{j=0}^{q} \int_0^{+\infty} |D_n^j u(x_n)|^2 dx_n + \sum_{j=0}^{q-1} |D_n^j u(0)|^2 \leq K \int_0^{+\infty} |P^-(0, \zeta', D_n)u(x_n)|^2 dx_n$$

Now identically as before we obtain the proof of (1.19).

Notice finally that from the proofs given above it follows that the cone C_o' may be an arbitrary cone in Z^{n+1} for which $P^+(x, \zeta', \zeta_n)$ and $P^-(x, \zeta', \zeta_n)$ have the property (a) in B.) .

E.) In this point we shall prove that

$$(1.23) \quad |\eta| \int_{E_+^{n+1}} |\psi u|^2 dx \leq K\Big(\int_{E_+^{n+1}} |M_j^+ \psi_1(x, D+i\eta)\psi u|^2 dx + \int_{R_o^n} |\psi u(x', 0)|^2 dx' , \quad j = 1, \ldots, r ,$$

$$(1.24) \quad |\eta| \int_{E_+^{n+1}} |\psi u|^2 dx + \int_{R_o^n} |\psi u(x', 0)|^2 dx' \leq$$

$$\leq K \int_{E_+^{n+1}} |M_k^- \psi_1(x, D+i\eta)\psi u|^2 dx , \quad k = 1, \ldots, s ,$$

hold for $u \in C^{\infty}_{(o)}(E_+^{n+1})$, $|\eta|$ sufficiently large, with some constant K independent of u and η .

Consider the expression

$$2\,\mathrm{Im}\int_{E_+^{n+1}} M_j^+\psi_1(x,D+i\eta)\psi u\overline{\psi u}\,dx = \int_{R_0^n}|\psi u(x',0)|^2\,dx' - \tag{1.25}$$

$$-\frac{1}{i}\int_{E_+^{n+1}}\left(\lambda_j^+\psi_1(x,D'+i\eta') - (\lambda_j^+\psi_1)^*(x,D'+i\eta')\right)\psi u\overline{\psi u}\,dx .$$

By the proposition (ii) the operator $-i(\lambda_j^+\psi_1(x,D'+i\eta') - (\lambda_j^+\psi_1)^*(x,D'+i\eta'))$ can be written as $2(\mathrm{Im}\,\lambda_j^+)\psi_1(x,D'+i\eta')+ S_0$, where S_0 is an operator of order 0. $\mathrm{Im}\,\lambda_j^+(x,\zeta') = 0$ for $\eta_0 = 0$, therefore $\dfrac{\mathrm{Im}\,\lambda_j^+(x,\zeta')}{\eta_0}$ can be uniquely prolonged to a homogeneous function of degree 0 from $C^\infty(\overline{E_+^{n+1}}\times C_0')$. By the Cauchy-Riemann equations

$$\lim_{\eta_0\to 0}\frac{\mathrm{Im}\,\lambda_j^+(x,\zeta')}{\eta_0} = \frac{\partial\,\mathrm{Im}\,\lambda_j^+(x,\xi')}{\partial\eta_0} = \frac{\partial\,\mathrm{Re}\,\lambda_j^+(x,\xi')}{\partial\xi_0} = \frac{\partial\lambda_j^+(x,\xi')}{\partial\xi_0}$$

and since $\lambda_j^+(x,\xi')$ is a simple root of the equation $P^0(x,\xi',\zeta_n) = 0$ we have

$$\frac{\partial\lambda_j^+(x,\xi')}{\partial\xi_0} = -\frac{\dfrac{\partial P^0(x,\xi',\lambda_j^+(x,\xi'))}{\partial\xi_0}}{\dfrac{\partial P^0(x,\xi',\lambda_j^+(x,\xi'))}{\partial\xi_n}} \neq 0 .$$

For sufficiently small ε we have

$$\frac{\operatorname{Im} \lambda_j^+(x, \zeta')}{\eta_0} \neq 0 \quad \text{for} \quad (x, \zeta') \in \overline{E_+^{n+1}} \times (C_0' \cap Z_-^{n+1}).$$

Define the function $\mu_j^+(x, \zeta')$:

$$\mu_j^+(x, \zeta') = \frac{2 \operatorname{Im} \lambda_j^+(x, \zeta')}{|\eta_0|} \psi_1(\zeta') \quad \text{for} \quad \eta_0 < 0,$$

$$\mu_j^+(x, \xi') = -2 \frac{\partial \lambda_j^+(x, \xi')}{\partial \xi_0} \psi_1(\xi').$$

Then $\mu_j^+ \in C^\infty(\overline{E_+^{n+1}} \times (\overline{Z_-^{n+1}} \setminus \{0\}))$, is homogeneous of degree 0 with respect to ζ' and there is a constant $\gamma_0 > 0$ such that

$$\mu_j^+(x, \zeta') \geqslant 2\gamma_0 \quad \text{for} \quad (x, \zeta') \in \overline{E_+^{n+1}} \times ((\operatorname{supp} \psi) \setminus \{0\}).$$

The proposition (iii) implies

$$\operatorname{Re} \int_{E_+^{n+1}} 2(\operatorname{Im} \lambda_j^+) \psi_1(x, D' + i\eta') \psi u \overline{\psi u} \, dx \geqslant \gamma_0 |\eta| \int_{E_+^{n+1}} |\psi u|^2 \, dx$$

for $|\eta|$ sufficiently large. This and (1.25) gives

$$\gamma_0 |\eta| \int_{E_+^{n+1}} |\psi u|^2 dx \leq \int_{E_+^{n+1}} |M_j^+ \psi_1(x, D + i\eta) \psi u|^2 dx + \int_{R_0^n} |\psi u(x', 0)|^2 dx' +$$

$$+ K \int_{E_+^{n+1}} |\psi u|^2 dx .$$

The inequality (1.23) easily follows from the above one.

The proof of (1.24) is similar. Considering the expression (1.25) with $M_k^- \psi_1$ instead of $M_j^+ \psi_1$ and defining the function $\mu_k^-(x, \zeta')$ as $- \frac{2 \operatorname{Im} \lambda_k^-(x, \zeta')}{|\eta_0|} \psi_1(\zeta')$ for $\eta_0 < 0$ and $2 \frac{\partial \lambda_k^-(x, \xi')}{\partial \xi_0} \psi_1(\xi')$ for $\eta_0 = 0$, we obtain

$$(1.26) \quad |\eta| \operatorname{Re} \int_{E_+^{n+1}} \mu_k^-(x, D' + i\eta') \psi u \overline{\psi u}\, dx + \int_{R_0^n} |\psi u(x', 0)|^2 dx' =$$

$$= 2 \operatorname{Im} \int_{E_+^{n+1}} M_k^- \psi_1(x, D+i\eta) \psi u \overline{\psi u}\, dx - \operatorname{Re} \int_{E_+^{n+1}} S_0 \psi u \overline{\psi u}\, dx .$$

Now the function $\mu_k^- \in C^\infty(\overline{E_+^{n+1}} \times (\overline{Z_-^{n+1}} \setminus \{0\}))$, is homogeneous of degree 0 with respect to ζ' and there is a constant $\gamma_0 > 0$ such that

$$\mu_k^-(x, \zeta') \geq 2\gamma_0 \qquad \text{for } (x, \zeta') \in \overline{E_+^{n+1}} \times ((\operatorname{supp} \psi') \setminus \{0\}).$$

The operator S_0 is of order 0. Applying the proposition (iii) again and estimating in the obvious manner the terms occuring in (1.26) we get (1.24).

F.) Now we are going to make the crucial step in our considerations. The following proposition holds for the operators $N_l \psi_1(x, D+i\eta)$, $l = 1, \ldots, t$.

There exist: a quadratic hermitian form $\sum_{\mu,\nu=0}^{1} p_{\mu,\nu}^{(1)}(\tau;\zeta)\zeta_n^\mu\overline{\zeta_n^\nu}$ of the variables η_0, ξ' and ζ_n with coefficients depending on a parameter τ, a constant $\tau_0 > 0$, a constant $\gamma_0(\tau) > 0$ for $\tau > \tau_0$ and a constant $\delta_0(\tau, \varepsilon_1) > 0$ for $\varepsilon_1 > 0$ and $\tau > \tau_0$, such, that for each $\varepsilon_1 > 0$

$$(1.27)\quad \gamma_0(\tau)|\eta|\,\|\psi u\|^2_{(1,0),\eta} + \int_{R_0^n}\sum_{\mu,\nu=0}^{1} p_{\mu,\nu}^{(1)}(\tau; D'+i\eta')D_n^\mu\psi u(x',0)\overline{D_n^\nu\psi u(x',0)}dx' - \varepsilon_1\sum_{\mu=0}^{1}\|D_n^\mu\psi u(\cdot,0)\|^2_{(1-\mu),\eta} \le \int_{E_+^{n+1}}|N_1\psi_1(x,D+i\eta)\psi u|^2dx,$$

$u \in C^\infty_{(0)}(E_+^{n+1})$,

for suitably chosen δ, $\varepsilon < \delta_0(\tau,\varepsilon_1)$ and for $|\eta|$ sufficiently large.

Moreover, the quadratic form $\sum_{\mu,\nu=0}^{1} p_{\mu,\nu}^{(1)}(\tau;\zeta')\zeta_n^\mu\overline{\zeta_n^\nu}$ satisfies

$$(1.28)\quad \sum_{\mu,\nu=0}^{1} p_{\mu,\nu}^{(1)}(\tau;\xi^{0\prime})\zeta_n^\mu\overline{\zeta_n^\nu} = |\zeta_n - \chi_{n,1}(0,\xi^{0\prime})|^2 + 2\frac{\gamma_1}{\tau}\left(\zeta_n + \overline{\zeta_n} - 2\chi_{n,1}(0,\xi^{0\prime})\right)$$

Consider operators $A(x,D+i\eta)$ and $B(x,D+i\eta)$ defined on $\overline{E_+^{n+1}}$, of the form

$$A(x,D+i\eta) = D_n^m + \sum_{j=1}^{m} a_j(x,D'+i\eta')D_n^{m-j},$$

$$B(x,D+i\eta) = D_n^{m-1} + \sum_{j=1}^{m-1} b_j(x,D'+i\eta')D_n^{m-1-j},$$

where $a_j(x,D'+i\eta')$, $b_j(x',D'+i\eta')$ are pseudodifferential operators of order j with real valued symbols for $\eta_0 = 0$. Then

$$(1.29)\quad 2\operatorname{Im}A(x,\zeta',\zeta_n)\overline{B(x,\zeta',\zeta_n)} =$$

$$= -i(\zeta_n - \overline{\zeta}_n)\sum_{\mu,\nu=0}^{m-1} p_{\mu,\nu}(x,\zeta')\zeta_n^{\mu}\overline{\zeta_n^{\nu}} +$$

$$+ |\eta|\sum_{\mu,\nu=0}^{m-1} q_{\mu,\nu}(x,\zeta')\zeta_n^{\mu}\overline{\zeta_n^{\nu}},$$

where $p_{\mu,\nu}$, $q_{\mu,\nu}$ are smooth and homogeneous functions of degree $2(m-1)-\mu-\nu$ and the quadratic forms

$$\sum_{\mu,\nu=0}^{m-1} p_{\mu,\nu}(x,\zeta')z_\mu\overline{z_\nu} \quad \text{and} \quad \sum_{\mu,\nu=0}^{m-1} q_{\mu,\nu}(x,\zeta')z_\mu\overline{z}_\nu,$$

$z = (z_0,z_1,\ldots,z_{m-1}) \in \mathbb{C}^m$, are hermitian.

To see it consider the expression $2\operatorname{Im}a(x,\zeta')\zeta_n^k\overline{b(x,\zeta')\zeta_n^l}$. Introduce the symbol $\equiv$ which means that the difference of terms standing on its sides has the form in the right hand side of the equation (1.29).

Because $2\operatorname{Im}a(x,\zeta')\zeta_n^k\overline{b(x,\zeta')\zeta_n^l} = -2\operatorname{Im}b(x,\zeta')\zeta_n^l\overline{a(x,\zeta')\zeta_n^k}$ we can assume that $k \geqslant l$ and write $k = l + p$. Then:

$$2\operatorname{Im}a\zeta_n^k\overline{b\zeta_n^l} = \zeta_n^l\overline{\zeta_n^l}\,2\operatorname{Im}a\zeta_n^p\overline{b} = (\operatorname{Re}a\operatorname{Re}b+\operatorname{Im}a\operatorname{Im}b)\zeta_n^l\overline{\zeta_n^l}\,2\operatorname{Im}\zeta_n^p +$$

$$+(\operatorname{Im}a\operatorname{Re}b - \operatorname{Re}a\operatorname{Im}b)\zeta_n^l\overline{\zeta_n^l}\,2\operatorname{Re}\zeta_n^p = -i(\zeta_n-\overline{\zeta}_n)(\operatorname{Re}a\operatorname{Re}b + \operatorname{Im}a\operatorname{Im}b)\cdot$$

$$\cdot(\zeta_n^{p-1} + \zeta_n^{p-2}\overline{\zeta}_n + \ldots + \zeta_n\overline{\zeta_n^{p-2}} + \overline{\zeta_n^{p-1}})\zeta_n^l\overline{\zeta_n^l} +$$

$$+ (\operatorname{Im}a\operatorname{Re}b - \operatorname{Re}a\operatorname{Im}b)\zeta_n^l\overline{\zeta_n^l}\cdot 2\operatorname{Re}\zeta_n^p \equiv (\operatorname{Im}a\operatorname{Re}b - \operatorname{Re}a\operatorname{Im}b)\zeta_n^l\overline{\zeta_n^l}\,2\operatorname{Re}\zeta_n^p.$$

If $p = 0$, then by the assumptions made on the functions a and b we have

$$2(\operatorname{Im}a\operatorname{Re}b - \operatorname{Re}a\operatorname{Im}b)\zeta_n^l\overline{\zeta_n^l} = |\eta|\Big(\frac{\operatorname{Im}a(x,\zeta')}{|\eta|}\operatorname{Re}b(x,\zeta') -$$

$$- \operatorname{Re}a(x,\zeta')\frac{\operatorname{Im}b(x,\zeta')}{|\eta|}\Big)\, 2\zeta_n^l\overline{\zeta_n^l} \equiv 0 .$$

If $p = 1$, then $l < m-1$ and again

$$(\operatorname{Im}a\operatorname{Re}b - \operatorname{Re}a\operatorname{Im}b)\zeta_n^l\overline{\zeta_n^l}\, 2\operatorname{Re}\zeta_n = |\eta|\frac{\operatorname{Im}a(x,\zeta')}{|\eta|}\operatorname{Re}b(x,\zeta') -$$

$$- \operatorname{Re}a(x,\zeta')\frac{\operatorname{Im}b(x,\zeta')}{|\eta|}\zeta_n^l\overline{\zeta_n^l}(\zeta_n + \overline{\zeta}_n) \equiv 0.$$

If $p \geq 2$, then $l < m-1$ also and

$$(\operatorname{Im}a\operatorname{Re}b - \operatorname{Re}a\operatorname{Im}b)\zeta_n^l\overline{\zeta_n^l}\, 2\operatorname{Re}\zeta_n^p =$$

$$= (\operatorname{Im}a\operatorname{Re}b - \operatorname{Re}a\operatorname{Im}b)\zeta_n^l\overline{\zeta_n^l}\, 2\operatorname{Re}(\zeta_n^p - \zeta_n^{p-1}\overline{\zeta}_n + \zeta_n^{p-1}\overline{\zeta}_n) =$$

$$= -i(\zeta_n - \overline{\zeta}_n)(\operatorname{Im}a\operatorname{Re}b - \operatorname{Re}a\operatorname{Im}b)(-2\operatorname{Im}\zeta_n^{p-1}) +$$

$$+ (\operatorname{Im}a\operatorname{Re}b - \operatorname{Re}a\operatorname{Im}b)\zeta_n^{l+1}\overline{\zeta_n^{l+1}}\, 2\operatorname{Re}\zeta_n^{p-2} \equiv$$

$$\equiv |\eta|\Big(\frac{\operatorname{Im}a(x,\zeta')}{|\eta|}\operatorname{Re}b(x,\zeta') - \operatorname{Re}a(x,\zeta')\frac{\operatorname{Im}b(x,\zeta')}{|\eta|}\Big)\zeta_n^{l+1}\overline{\zeta_n^{l+1}}(\zeta_n^{p-2} + \overline{\zeta_n^{p-2}}) \equiv$$

$$\equiv 0 .$$

Thus the equality (1.29) is proved. Integrating by parts the expression $2\operatorname{Im}\int_{E_+^{n+1}} A(x,D+i\eta)u\overline{B(x,D+i\eta)u}\, dx$ and using the calculus of pseudodifferential operators (the propositions (i), (ii)) we obtain

$$(1.30)\qquad 2\,\mathfrak{Im}\int_{E_+^{n+1}} A(x,D+i\eta)u\overline{B(x,D+i\eta)u}\,dx =$$

$$= |\eta|\ \mathrm{Re}\int_{E_+^{n+1}} \sum_{\mu,\nu=0}^{m-1} q_{\mu,\nu}(x,D'+i\eta')D_n^{\mu}u\overline{D_n^{\nu}u}\,dx\ +$$

$$+\ \mathrm{Re}\int_{R_0^n} \sum_{\mu,\nu=0}^{m-1} p_{\mu,\nu}(x',D'+i\eta')D_n^{\mu}u(x',0)\overline{D_n^{\nu}u(x',0)}\,dx'\ +$$

$$+\ K(u)\ ,\qquad u\in C_{(0)}^{\infty}(E_+^{n+1})\ ,$$

where

$$|K(u)| \le K\|u\|^2_{(m-1,0),\eta}\ .$$

Writing in this way we understand that the operators $q_{\mu,\nu}(x,D'+i\eta')$ act only on the function $D_n^{\mu}u$, not on $\overline{D_n^{\nu}u}$. The same remark applies to $p_{\mu,\nu}(x',D'+i\eta')$ and in general to all quadratic pseudodifferential forms occuring in the future.

From the proof of the equality (1.29) it follows also that if the term D_n^m does not occur in $A(x,D+i\eta)$, then the estimations

$$(1.31)\qquad \sup_{x',|\zeta'|=1}|p_{\mu,\nu}(x',\zeta')| \le c\sup_{j,|\zeta'|=1}|a_j(x',\zeta')|\sup_{j,x',|\zeta'|=1}(|b_j(x',\zeta')|,1)\ ,$$

$$(1.32)\qquad \sup_{x,|\zeta'|=1}|q_{\mu,\nu}(x,\zeta')| \le c\sup_{j,x,|\zeta'|=1}\left(|a_j(x,\zeta')|,\left|\frac{\mathfrak{Im}\,a_j(x,\zeta')}{|\eta_0|}\right|\right)\cdot$$

$$\cdot\sup_{j,x,|\zeta'|=1}\left(|b_j(x,\zeta')|,\left|\frac{\mathfrak{Im}\,b_j(x,\zeta')}{|\eta_0|}\right|,1\right)$$

hold with a constant c dependent on m only.

Now we shall apply the above considerations to the expression

$$(1.33)\qquad 2\,\mathrm{Im}\int_{E_+^{n+1}} N_1\psi_1(x,D+i\eta)\psi u\overline{L(D+i\eta)\psi u}\,dx\ ,$$

where $L(D+i\eta)$ is a differential operator of the first order. To simplify notations we shall omit the index 1. Take the operator

$$N'(x,D+i\eta) = N\psi_1(x,D+i\eta) - N\psi_1(0,D+i\eta) =$$
$$= (p\psi_1(x,D'+i\eta') - p\psi_1(0,D'+i\eta'))D_n +$$
$$+ (q\psi_1(x,D'+i\eta') - q\psi_1(0,D'+i\eta'))$$

instead of $N_1\psi_1(x,D+i\eta)$ in (1.33). Then

$$(1.34)\qquad 2\,\mathrm{Im}\int_{E_+^{n+1}} N'(x,D+i\eta)\psi u\,\overline{L(D+i\eta)\psi u}\,dx =$$
$$= |\eta|\ \mathrm{Re}\int_{E_+^{n+1}}\sum_{\mu,\nu=0}^{1} q'(x,D'+i\eta')D_n^{\mu}\psi u\overline{D_n^{\nu}\psi u}\,dx +$$
$$+ \mathrm{Re}\int_{R^n}\sum_{\mu,\nu=0}^{1} p'_{\mu,\nu}(x',D'+i\eta')D_n^{\mu}\psi u(x',0)\overline{D_n^{\nu}\psi u(x',0)}\,dx' +$$
$$+\ K'(\psi u)\ ,$$

$$u\in C_{(0)}^{\infty}(E_+^{n+1})\ ,\qquad |K'(\psi u)|\le K'\|u\|^2_{(1,0),\eta}\ .$$

(1.31), (1.32) implies that

$$(1.35)\quad \sup_{x',|\zeta'|=1} |p'_{\mu,\nu}(x',\zeta')| \leq$$

$$\leq K_1 \max\Big(\sup_{x',|\zeta'|=1} \big|\big(p(x',\zeta')-p(0,\zeta')\big)\psi_1(\zeta')\big| \,,\ \sup_{x',|\zeta'|=1} \big|\big(q(x',\zeta')-q(0,\zeta')\big)\psi_1(\zeta')\big|\Big)$$

$$(1.36)\quad \sup_{x,|\zeta'|=1} |q'_{\mu,\nu}(x,\zeta')| \leq K_1 \max\Big(\sup_{x,|\zeta'|=1} \big|(p(x,\zeta') - p(0,\zeta'))\psi_1(\zeta')\big|,$$

$$\sup_{x,|\zeta'|=1} \big|(q(x,\zeta') - q(0,\zeta'))\psi_1(\zeta')\big|,\ \sup_{x,|\zeta'|=1} \Big|\frac{\Im m(p(x,\zeta')-p(0,\zeta'))}{|\eta_0|}\psi_1(\zeta')\Big|,$$

$$\sup_{x,|\zeta'|=1} \Big|\frac{\Im m(q(x,\zeta') - q(0,\zeta'))}{|\eta_0|}\psi_1(\zeta')\Big|\Big),$$

where the constant K_1 depends only on the coefficients of the operator L, as it is seen from (1.31), (1.32). In the future L will have coefficients depending on a parameter τ and then $K_1 = O(\tau)$ for $\tau \longrightarrow +\infty$.

In view of (1.35), (1.36)

$$\sup_{x',|\zeta'|=1} |p'_{\mu,\nu}(x',\zeta')| = O(\delta)\ ,\quad \sup_{x,|\zeta'|=1} |q'_{\mu,\nu}(x,\zeta')| = O(\delta)\ ,$$

so it follows from the proposition (iv) that

$$(1.37)\quad \Big|\int_{E_+^{n+1}} \sum_{\mu,\nu=0}^{1} q'_{\mu,\nu}(x,D'+i\eta')D_n^{\mu}\psi u\overline{D_n^{\nu}\psi u}\,dx\Big| \leq O(\delta)\|\psi u\|^2_{(1,0),\eta}\ ,$$

$$(1.38)\quad \Big|\int_{R_0^n} \sum_{\mu,\nu=0}^{1} p'_{\mu,\nu}(x',D'+i\eta')D_n^{\mu}\psi u(x',0)\overline{D_n^{\nu}\psi u(x',0)}\,dx'\Big| \leq$$

$$\leq O(\delta)\sum_{\mu=0}^{1} \|D_n^{\mu}\psi u(\cdot,0)\|^2_{(1-\mu),\eta}$$

for $u \in C^{\infty}_{(0)}(E^{n+1}_{+})$ and $|\eta|$ larger than a certain number dependent on δ, ε and K_1.

Thus, by the definition of $N'(x, D+i\eta)$, (1.34), (1.37) and (1.38) for each $\varepsilon_1 > 0$ there exists a constant $\delta_0(\varepsilon_1, K_1)$ such that

$$(1.39) \quad 2\,\mathrm{Im}\int_{E^{n+1}_{+}} N\psi_1(x, D+i\eta)\psi u\overline{L(D+i\eta)\psi u}\,dx \geqslant$$

$$\geqslant 2\,\mathrm{Im}\int_{E^{n+1}_{+}} N\psi_1(0, D+i\eta)\psi u\overline{L(D+i\eta)\psi u}\,dx -$$

$$-\varepsilon_1|\eta|\,\|\psi u\|^2_{(1,0),\eta} - \varepsilon_1\sum_{\mu=0}^{1}\|D_n^{\mu}\psi u(\cdot,0)\|^2_{(1-\mu),\eta} -$$

$$- K\,\|\psi u\|^2_{(1,0),\eta}$$

for all $u \in C^{\infty}_{(0)}(E^{n+1}_{+})$, $0 < \delta < \delta_0(\varepsilon_1, K_1)$, $|\eta| > T_1(\varepsilon, \delta, K_1)$.

The next step is to replace the operator $N\psi_1(0, D+i\eta)$ in the integral on the right hand side of (1.39) by a suitable differential operator of the second order.
To do so we must carefully investigate the symbol $N(0, \zeta', \zeta_n)$ of the operator $N\psi_1(0, D+i\eta)$ for real values of the arguments. Consider the function

$$N(0, \xi', \xi_n) = \xi_n^2 + p(0, \xi')\xi_n + q(0, \xi')$$

on the hyperplane $\langle \xi, \xi^{0\prime\prime}\rangle = 1$. The equation $\langle \xi, \xi^{0\prime\prime}\rangle = 1$ is equivalent to $\langle \xi'', \xi^{0\prime\prime}\rangle = 1$ and the coordinates ξ_0, ξ_n of points in this hyperplane can admit arbitrary values.

Introduce new variables ϑ in the hyperplane. Let us write $\vartheta = (\vartheta_1, \vartheta_2, \ldots, \vartheta_{n-1}, \vartheta_n)$, $\vartheta' = (\vartheta_1, \vartheta_2, \ldots, \vartheta_{n-1})$, $\vartheta'' = (\vartheta_2, \ldots, \vartheta_{n-1})$. Let ϑ'' give some parametrization of the hyperplane $\langle \xi'', \xi^{0''} \rangle = 1$ in R^{n-1}, e.g. there is a rectangular matrix A' with $n-1$ rows and $n-2$ columns such that $\langle \xi'', \xi^{0''} \rangle = 1$ if and only if $\xi'' = A'\vartheta'' + \xi^{0''}$ for some $\vartheta'' \in R^{n-2}$. Hence $\langle A'\vartheta'', \xi^{0''} \rangle = 0$ for $\vartheta'' \in R^{n-2}$. For further purposes we choose the matrix A' in such way that its columns form an orthonormal basis in the space $\langle \xi'', \xi^{0''} \rangle = 0$. Then ${}^tA'A' = I$, where ${}^tA'$ denotes the transposed matrix, and if $\xi'' = A'\vartheta'' + \xi^{0''}$ then $\vartheta'' = {}^tA'(\xi'' - \xi^{0''})$ and

$$(1.40) \quad |\vartheta''|^2 = \langle {}^tA'(\xi'' - \xi^{0''}), \ {}^tA'(\xi'' - \xi^{0''}) \rangle = |\xi'' - \xi^{0''}|^2 .$$

Further, let

$$\xi_0 = \chi_0(0, \xi^{0''}) + \langle \operatorname{grad}_{\vartheta''} \chi_0(0, A'\vartheta'' + \xi^{0''})\big|_{\vartheta''=0}, \vartheta'' \rangle + \vartheta_1 ,$$

or $\xi_0 = \xi_0^0 + \langle \operatorname{grad}_{\xi''} \chi_0(0, \xi^{0''}), A'\vartheta'' \rangle + \vartheta_1$.

We write $\xi_0^0 = \chi_0(0, \xi^{0''})$, $\xi_n^0 = \chi_n(0, \xi^{0''})$. From the above equality we have

$$\vartheta_1 = \xi_0 - \xi_0^0 - \langle \operatorname{grad}_{\xi''} \chi_0(0, \xi^{0''}), \xi'' - \xi^{0''} \rangle =$$

$$= \xi_0 - \langle \operatorname{grad}_{\xi''} \chi_0(0, \xi^{0''}), \xi'' \rangle \quad \text{for} \quad \langle \xi'', \xi^{0''} \rangle = 1.$$

The variables ξ' and ϑ' are connected by the equation

$$\xi' = A\vartheta' + \xi^{0\prime}, \quad \langle \xi', \xi^{0\prime\prime} \rangle = 1, \quad \vartheta' \in R^{n-1}.$$

Finally let us introduce ϑ_n by

$$\xi_n = \vartheta_n + \xi^0_n .$$

In the new variables the function $N(0, \xi', \xi_n)$ has the form

$$(1.41) \qquad (\vartheta_n + \xi^0_n)^2 + p(0, A\vartheta' + \xi^{0\prime})(\vartheta_n + \xi^0_n) + q(0, A\vartheta' + \xi^{0\prime}) =$$

$$= \vartheta_n^2 + (p(0, A\vartheta' + \xi^{0\prime}) + 2\xi^0_n)\vartheta_n + (q(0, A\vartheta' + \xi^{0\prime}) +$$

$$+ p(0, A\vartheta' + \xi^{0\prime})\xi^0_n + (\xi^0_n)^2) \stackrel{\text{def}}{=} \vartheta_n^2 + p_1(\vartheta')\vartheta_n + q_1(\vartheta'),$$

where p_1 and q_1 are smooth functions defined on some neighbourhood of 0.

The hyperplane $\vartheta_1 = 0$ is tangent to the hypersurface $\xi_0 = \chi_0(0, \xi'')$ at the point $\vartheta' = 0$, therefore in the new variables this hypersurface is defined by the equation $\vartheta_1 = \Theta_0(\vartheta'')$, where Θ_0 is a smooth function defined on some neighbourhood of 0 and

$$\Theta_0(0) = 0, \qquad \frac{\partial \Theta_0(0)}{\partial \vartheta_i} = 0, \quad i = 2, \ldots, n-1.$$

By the definition of χ_0 the set of points of the hypersurface $\vartheta_1 = \Theta_0(\vartheta'')$ coincides with the set of points ϑ' for which the equation $\vartheta_n^2 + p_1(\vartheta')\vartheta_n + q_1(\vartheta') = 0$ in ϑ_n has a double root. Denote this root by $\Theta_n(\vartheta'')$, it is the function $\chi_n(0, \xi'')$ in the new variables. Thus we have

$$\vartheta_n^2 + p_1(\Theta_0(\vartheta''),\vartheta'')\vartheta_n + q_1(\Theta_0(\vartheta''),\vartheta'') = (\vartheta_n - \Theta_n(\vartheta''))^2$$

and $p_1(\Theta_0(\vartheta''),\vartheta'') = -2\Theta_n(\vartheta''),\quad q_1(\Theta_0(\vartheta''),\vartheta'') = \Theta_n^2(\vartheta'')$.

The above equalities and $\Theta_n(0) = 0$ imply

$$p_1(0) = 0,\quad q_1(0) = 0\ ,\quad \frac{\partial q_1(0)}{\partial \vartheta_i} = 0\ ,\quad i = 2,\dots,n-1.$$

Let us introduce the Taylor's expansion to the first order at the point 0 for the functions $p_1(\vartheta')$ and $q_1(\vartheta')$ in (1.41). We obtain the function

$$\vartheta_n^2 + p_1(0)\vartheta_n + \sum_{i=1}^{n-1}\frac{\partial p_1(0)}{\partial \vartheta_i}\vartheta_i\vartheta_n + q_1(0) + \sum_{i=1}^{n-1}\frac{\partial q_1(0)}{\partial \vartheta_i}\vartheta_i =$$

$$= \vartheta_n^2 + \sum_{i=1}^{n-1}\frac{\partial p_1(0)}{\partial \vartheta_i}\vartheta_i\vartheta_n + \frac{\partial q_1(0)}{\partial \vartheta_1}\vartheta_1 .$$

Adding to it the quadratic form in ϑ' $\frac{1}{4}\left(\sum_{i=1}^{n-1}\frac{\partial p_1(0)}{\partial \vartheta_i}\vartheta_i\right)^2 - \tau^2\vartheta_1^2 + |\vartheta''|^2$ we get finally the function

$$(1.42)\qquad \left(\vartheta_n + \frac{1}{2}\sum_{i=1}^{n-1}\frac{\partial p_1(0)}{\partial \vartheta_i}\vartheta_i\right)^2 + \frac{\partial q_1(0)}{\partial \vartheta_1}\vartheta_1 - \tau^2\vartheta_1^2 + |\vartheta''|^2.$$

The difference of (1.41) and (1.42) has the form

$$p_2(\vartheta')\vartheta_n + q_2(\vartheta')$$

and $p_2(\vartheta') = O(|\vartheta'|^2)$, $q_2(\vartheta') = O(|\vartheta'|^2)$ in a neighbourhood of 0.

Let us write the function (1.42) in the old variables.

By (1.40), (1.41) and (1.42) it is

$$(1.43)\qquad (\xi_n - \xi_n^o + \tfrac{1}{2}\langle \operatorname{grad}_{\vartheta'} p(0, A\vartheta' + \xi^{o\prime})\big|_{\vartheta'=0}, \vartheta'\rangle)^2 +$$

$$+\langle \operatorname{grad}_{\vartheta'} q(0, A\vartheta' + \xi^{o\prime})\big|_{\vartheta'=0}, \vartheta'\rangle + \langle \operatorname{grad}_{\vartheta'} p(0, A\vartheta' + \xi^{o\prime})\big|_{\vartheta'=0}, \vartheta'\rangle \xi_n^o -$$

$$-\tau^2(\xi_o - \langle \operatorname{grad}_{\xi''} \chi_o(0, \xi^{o\prime\prime}), \xi''\rangle)^2 + |\xi'' - \xi^{o\prime\prime}|^2 =$$

$$= (\xi_n + \tfrac{1}{2}\langle \operatorname{grad}_{\xi'} p(0, \xi^{o\prime}), \xi'\rangle)^2 + \langle \operatorname{grad}_{\xi'} q(0, \xi^{o\prime}), \xi'\rangle +$$

$$+\langle \operatorname{grad}_{\xi'} p(0, \xi^{o\prime}), \xi'\rangle \xi_n^o - \tau^2(\xi_o - \langle a, \xi''\rangle)^2 + |\xi'' - \xi^{o\prime\prime}|^2$$

for $\langle \xi'', \xi^{o\prime\prime}\rangle = 1$, where $a = \operatorname{grad}_{\xi''} \chi_o(0, \xi^{o\prime\prime})$.
(1.43) can be uniquely prolonged on the whole space R^{n+1} to the quadratic form:

$$(1.44)\qquad K(\xi) = \xi_n^2 + p_o(\xi')\xi_n + q_o(\xi') =$$

$$= \xi_n^2 + \langle \operatorname{grad}_{\xi'} p(0, \xi^{o\prime}), \xi'\rangle \xi_n + \tfrac{1}{4}(\langle \operatorname{grad}_{\xi'} p(0, \xi^{o\prime}), \xi'\rangle)^2 +$$

$$+\langle \operatorname{grad}_{\xi'} q(0, \xi^{o\prime}), \xi'\rangle \langle \xi'', \xi^{o\prime\prime}\rangle +$$

$$+\langle \operatorname{grad}_{\xi'} p(0, \xi^{o\prime}), \xi'\rangle \xi_n^o \langle \xi'', \xi^{o\prime\prime}\rangle - \tau^2(\xi_o - \langle a, \xi''\rangle)^2 +$$

$$+ |\xi'' - \xi^{o\prime\prime}\langle \xi'', \xi^{o\prime\prime}\rangle|^2 .$$

The forms p_o and q_o are defined for all complex values of arguments. For $\langle \xi'', \xi^{o\prime\prime}\rangle = 1$ the inequalities

$$(1.45)\qquad |p(0,\zeta') - p_o(\zeta')| \leq c_1|\zeta' - \xi^{o\prime}|^2 ,$$

$$|q(0,\zeta') - q_o(\zeta')| \leq c_1|\zeta' - \xi^{o\prime}|^2 ,$$

$$\left|\frac{\partial p(0,\zeta')}{\partial \xi_i} - \frac{\partial p_o(\zeta')}{\partial \xi_i}\right| \leq c_1|\zeta' - \xi^{o\prime}| ,$$

$$\left|\frac{\partial q(0,\zeta')}{\partial \xi_i} - \frac{\partial q_o(\zeta')}{\partial \xi_i}\right| \leq c_1|\zeta' - \xi^{o\prime}| ,$$

$i = 0,1,\dots,n-1$,

hold for ζ' in some neighbourhood of $\xi^{o\prime}$. From the homogeneity of p, p_o, q, q_o and from the estimation

$$\left|\frac{\zeta'}{\langle \xi'', \xi^{o''}\rangle} - \xi^{o\prime}\right| = \frac{1}{|\langle \xi'', \xi^{o''}\rangle|}\left|\zeta' - \xi^{o\prime}(1 - \langle \xi'', \xi^{o''}\rangle)\right| \leq$$

$$\leq \frac{1}{|\langle \xi'', \xi^{o''}\rangle|}(1 + |\xi^{o\prime}|)|\zeta' - \xi^{o\prime}|$$

it follows that (1.45) holds for ξ'' in some neighbourhood of $\xi^{o''}$, for example such that in it additionally $\frac{1}{2} < \langle \xi'', \xi^{o''}\rangle < 2$. Again using the homogeneity we obtain

$$(1.46)\qquad |p(0,\zeta') - p_o(\zeta')| \leq c_2|\zeta'|^{-1}\left|\zeta' - \frac{|\zeta'|}{|\xi^{o\prime}|}\xi^{o\prime}\right|^2 ,$$

$$|q(0,\zeta') - q_o(\zeta')| \leq c_2\left|\zeta' - \frac{|\zeta'|}{|\xi^{o\prime}|}\xi^{o\prime}\right|^2 ,$$

$$\left|\frac{\partial p(0,\zeta')}{\partial \xi_i} - \frac{\partial p_o(\zeta')}{\partial \xi_i}\right| \leq c_2|\zeta'|^{-1}\left|\zeta' - \frac{|\zeta'|}{|\xi^{o\prime}|}\xi^{o\prime}\right| ,$$

$i = 0,1,\dots,n-1$,

$$\left| \frac{\partial q(0,\zeta')}{\partial \xi_i} - \frac{\partial q_o(\zeta')}{\partial \xi_i} \right| \leq c_2 \left| \zeta' - \frac{|\zeta'|}{|\xi^{o\prime}|}\xi^{o\prime} \right| ,$$

$$i = 0,1,\ldots,n-1 ,$$

for $\zeta' \in C_o'$ with ε sufficiently small.

The constant c_2 in above depends on the parameter in q_o. Consider now the coefficient $\frac{\partial q_1(0)}{\partial \vartheta_1}$. By (1.41) and the conclusion of the point B.) :

$$q_1(\vartheta_1,0) = q(0,\vartheta_1+\xi_0^o,\xi^{o\prime\prime}) + p(0,\vartheta_1+\xi_0^o,\xi^{o\prime\prime})\xi_n^o + (\xi_n^o)^2 =$$

$$= (\xi_n^o - z_n^+(0,\vartheta_1+\xi_0^o,\xi^{o\prime\prime}))(\xi_n^o - z_n^-(0,\vartheta_1+\xi_0^o,\xi^{o\prime\prime})) =$$

$$= \varphi(0,\xi^{o\prime\prime}, \overline{+a_1(0,\xi^{o\prime\prime})\sqrt{\vartheta_1}})\,\varphi(0,\xi^{o\prime\prime}, \overline{-a_1(0,\xi^{o\prime\prime})\sqrt{\vartheta_1}}) ,$$

and from the properties of the function φ it is

$$q_1(\vartheta_1,0) = -a_1^2(0,\xi^{o\prime\prime})\vartheta_1 + \ldots$$

where the dots denote the higher order terms in ϑ_1. Hence $\frac{\partial q_1(0)}{\partial \vartheta_1} = -a_1^2(0,\xi^{o\prime\prime})$, and denoting $a_1^2(0,\xi^{o\prime\prime}) = 2\gamma_1$ we have $\frac{\partial q_1(0)}{\partial \vartheta_1} = -2\gamma_1$, γ_1 real and $\gamma_1 \neq 0$. The function (1.42) can be written in the following form:

$$(1.47) \qquad \left(\vartheta_n + \frac{1}{2}\sum_{i=1}^{n-1}\frac{\partial p_1(0)}{\partial \vartheta_i}\vartheta_i\right)^2 + |\vartheta''|^2 + \left(\frac{\gamma_1}{\tau}\right)^2 - \left(\tau\vartheta_1 + \frac{\gamma_1}{\tau}\right)^2.$$

Simultaneously with this function we shall consider the function

$$(1.48)\qquad \vartheta_n + \frac{1}{2}\sum_{i=1}^{n-1} \frac{\partial p_1(0)}{\partial \vartheta_i}\vartheta_i + 2\left(\tau\vartheta_1 + \frac{\gamma_1}{\tau}\right).$$

In the old variables ξ it has the form

$$\xi_n - \xi_n^o + \frac{1}{2}\langle \mathrm{grad}_{\xi'}\ p(0,\xi^{o\prime}),\ \xi' - \xi^{o\prime}\rangle + 2\tau(\xi_o - \langle a,\xi''\rangle) + 2\frac{\gamma_1}{\tau} =$$

$$= \xi_n + \frac{1}{2}\langle \mathrm{grad}_{\xi'}\, p(0,\xi^{o\prime}),\xi'\rangle + 2\tau\,(\xi_o - \langle a,\xi''\rangle) + 2\frac{\gamma_1}{\tau}$$

for $\langle \xi'', \xi^{o\prime\prime}\rangle = 1$ and it can be uniquely prolonged to the linear function defined on the whole space R^{n+1} :

$$L(\xi) = \xi_n + \gamma_o(\xi') = \xi_n + \frac{1}{2}\langle \mathrm{grad}_{\xi'}\, p(0,\xi^{o\prime}),\xi'\rangle +$$

$$+ 2\tau\,(\xi_o - \langle a,\xi''\rangle) + 2\frac{\gamma_1}{\tau}\langle \xi'', \xi^{o\prime\prime}\rangle \quad .$$

Now we shall prove that the equation $K(\xi) = 0$ has two real different roots with respect to ξ_o for $(\xi'', \xi_n) \neq 0$ and that the root of $L(\xi) = 0$ in ξ_o separate them.
At first we shall show it for ξ in the hyperplane $\langle \xi, \xi^{o\prime\prime}\rangle = 1$. Lines in this hyperplane parallel to the axis ξ_o are parallel to the axis ϑ_1 also, hence it is sufficient to prove that the root in ϑ_1 of the function (1.4 8) separates the roots of (1.47). Introduce new variables $\lambda = (\lambda_1, \dots, \lambda_n)$ by the substitution $\lambda' = \vartheta'$, $\lambda_n = \vartheta_n + \frac{1}{2}\sum_{i=1}^{n-1} \frac{\partial p_1(0)}{\partial \vartheta_i}\vartheta_i$

Lines parallel to the axis ϑ_1 are described by the following parametrical representation in the new variables:

(1.49) $\lambda_1 = \lambda_1\,,\quad \lambda'' = \lambda^{\circ\prime\prime}\,,\quad \lambda_n = c_1\lambda_1 + \lambda^\circ_n\,,$

where $\lambda^{\circ\prime\prime}$ and λ°_n are arbitrary, $c_1 = \frac{1}{2}\frac{\partial p_1(0)}{\partial \vartheta_1}$.

The functions (1.47) and (1.48) take the form

$$\lambda_n^2 + |\lambda''|^2 + \left(\frac{\gamma_1}{\tau}\right)^2 - \left(\tau\lambda_1 + \frac{\gamma_1}{\tau}\right)^2 \quad\text{and}\quad \lambda_n + 2\left(\tau\lambda_1 + \frac{\gamma_1}{\tau}\right),$$

and their roots are described by the equations:

$$\lambda_1 = -\frac{\gamma_1}{\tau^2} \pm \frac{1}{\tau}\sqrt{\lambda_n^2 + |\lambda''|^2 + \left(\frac{\gamma_1}{\tau}\right)^2} \quad\text{and}\quad \lambda_1 = -\frac{\gamma_1}{\tau^2} - \frac{1}{2\tau}\lambda_n\,.$$

Now geometrically it is evident that the surface described by the second equation separates the two surfaces described by the first one and that the lines (1.49) intersect them if the parameter τ is sufficiently large, say $\tau > \tau_0$. By the homogeneity of $K(\xi)$ and $L(\xi)$ it follows the desired statement for all points ξ out of the hyperplane $\langle \xi, \xi^{\circ\prime\prime}\rangle = \langle \xi'', \xi^{\circ\prime\prime}\rangle = 0$. For ξ in this hyperplane we have:

$$K(\xi) = \left(\xi_n + \tfrac{1}{2}\langle \operatorname{grad}_{\xi'} p(0,\xi^{\circ\prime}), \xi'\rangle\right)^2 + |\xi''|^2 - \tau^2\left(\xi_0 - \langle a, \xi''\rangle\right)^2$$

$$L(\xi) = \xi_n + \tfrac{1}{2}\langle \operatorname{grad}_{\xi'} p(0,\xi^{\circ\prime}), \xi'\rangle + 2\tau\left(\xi_0 - \langle a, \xi''\rangle\right),$$

thus the roots of these forms satisfy the equations:

$$\xi_0 = \langle a, \xi''\rangle \pm \frac{1}{\tau}\sqrt{\left(\xi_n + \tfrac{1}{2}\langle \operatorname{grad}_{\xi'} p(0,\xi^{\circ\prime}), \xi'\rangle\right)^2 + |\xi''|^2}\,,$$

$$\xi_0 = \langle a, \xi'' \rangle - \frac{1}{2\tau}\left(\xi_n + \tfrac{1}{2}\langle \operatorname{grad}_{\xi'} p(0,\xi^{0\prime}), \xi' \rangle \right),$$

and now the separation is evident. The following reasoning shows that the first equation has roots. Introduce the variables λ : $\lambda_0 = \xi_0 - \langle a, \xi'' \rangle$, $\lambda'' = \xi''$, $\lambda_n = \xi_n + \frac{1}{2}\langle \operatorname{grad}_{\xi'} p(0,\xi^{0\prime}), \xi' \rangle$. Then the form $K(\xi)$ can be written as $\lambda_n^2 + |\lambda''|^2 - \tau^2\lambda_0^2$ and it has the roots described by $\lambda_0 = \pm \frac{1}{\tau}\sqrt{\lambda_n^2 + |\lambda''|^2}$. Lines parallel to the axis ξ_0 have the equations $\lambda_0 = \lambda_0$, $\lambda'' = \lambda^{0\prime\prime}$, $\lambda_n = c_1\lambda_0 + \lambda_n^0$ and they intersect the surfaces $\lambda_0 = \pm \frac{1}{\tau}\sqrt{\lambda_n^2 + |\lambda''|^2}$ for large τ.

We have proved that $K(\xi)$ is strongly hyperbolic with respect to ξ_0 and the hyperplane $L(\xi) = 0$ separate the sheets of the characteristic cone $K(\xi) = 0$. Hence

$$(1.50) \qquad K(\zeta_0, \xi'', \xi_n) = a_0 \prod_{i=1}^{2} (\zeta_0 - \lambda_i(\xi'', \xi_n)),$$

$$(1.51) \qquad L(\zeta_0, \xi'', \xi_n) = b_0 \sum_{i=1}^{2} \mu_i(\xi'', \xi_n)(\zeta_0 - \lambda_i(\xi'', \xi_n)),$$

where $\mu_i(\xi'', \xi_n) > 0$ and $-a_0 b_0 > 0$.
At the point $\xi^{0\prime}$

$$(1.52) \qquad K(\xi^{0\prime}, \zeta_n) = (\zeta_n - \xi_n^0)^2, \quad L(\xi^{0\prime}, \zeta_n) = \zeta_n - \xi_n^0 + 2\frac{\gamma_1}{\tau}.$$

Consider now again the expression (1.33) but with the operator

$$N''(D + i\eta) = N\psi_1(0, D + i\eta) - K\psi_1(D + i\eta)$$

instead of $N_1\psi_1(x, D + i\eta)$. Then

$$(1.53) \qquad 2\,\mathrm{Im}\int_{E_+^{n+1}} N''(D+i\eta)\psi u\,\overline{L(D+i\eta)\psi u}\,dx =$$

$$= |\eta|\,\mathrm{Re}\int_{E_+^{n+1}} \sum_{\mu,\nu=0}^{1} q''_{\mu,\nu}(D'+i\eta')D_n^{\mu}\psi u\overline{D_n^{\nu}\psi u}\,dx \; +$$

$$+ \mathrm{Re}\int_{R_0^n} \sum_{\mu,\nu=0}^{1} p''_{\mu,\nu}(D'+i\eta')D_n^{\mu}\psi u(x',0)\overline{D_n^{\nu}\psi u(x',0)}\,dx' +$$

$$+ K''(\psi u) ,$$

for $u \in C_{(0)}^{\infty}(E_+^{n+1})$, $|K''(\psi u)| \leq K''\|\psi u\|^2_{(1,0),\eta}$.

The functions $p''_{\mu,\nu}(\zeta')$ and $q''_{\mu,\nu}(\zeta')$ satisfy

$$\sup_{|\zeta'|=1}|p''_{\mu,\nu}(\zeta')| \leq K_2\max\Big(\sup_{|\zeta'|=1}|(p(0,\zeta') - p_0(\zeta'))\psi_1(\zeta')| ,$$

$$, \sup_{|\zeta'|=1}|(q(0,\zeta') - q_0(\zeta'))\psi_1(\zeta')|\Big) ,$$

$$\sup_{|\zeta'|=1}|q''_{\mu,\nu}(\zeta')| \leq K_2\max\Big(\sup_{|\zeta'|=1}|(p(0,\zeta') - p_0(\zeta'))\psi_1(\zeta')| ,$$

$$, \sup_{|\zeta'|=1}|(q(0,\zeta') - q_0(\zeta'))\psi_1(\zeta')| ,$$

$$, \sup_{|\zeta'|=1}\left|\frac{\mathrm{Im}(p(0,\zeta') - p_0(\zeta'))}{|\eta_0|}\psi_1(\zeta')\right| ,$$

$$, \sup_{|\zeta'|=1}\left|\frac{\mathrm{Im}(q(0,\zeta') - q_0(\zeta'))}{|\eta_0|}\psi_1(\zeta')\right|\Big)$$

and the inequalities (1.46) imply that

$$\sup_{|\zeta'|=1} \left| p''_{\mu,\nu}(\zeta') \right| = O(\varepsilon), \quad \sup_{|\zeta'|=1} \left| q''_{\mu,\nu}(\zeta') \right| = O(\varepsilon)$$

for $\varepsilon \to 0^+$.

By the proposition (iv) it is

$$(1.54) \quad \left| \int_{E_+^{n+1}} \sum_{\mu,\nu=0}^{1} q''_{\mu,\nu}(D'+i\eta') D_n^{\mu} \psi u \overline{D_n^{\nu} \psi u} \, dx \right| \leq O(\varepsilon) \| \psi u \|^2_{(1,0),\eta}$$

$$(1.55) \quad \left| \int_{R_0^n} \sum_{\mu,\nu=0}^{1} p_{\mu,\nu}(D'+i\eta') D_n^{\mu} \psi u(x',0) \overline{D_n^{\nu} \psi u(x',0)} dx' \right| \leq$$

$$\leq O(\varepsilon) \sum_{\mu=0}^{1} \| D_n^{\mu} \psi u(\cdot,0) \|^2_{(1-\mu),\eta} \quad \text{for } u \in C^{\infty}_{(o)}(E_+^{n+1})$$

and $|\eta|$ larger than a certain number dependent on ε and K_2.

The definition of $N''(D+i\eta)$, (1.53), (1.54) and (1.55) imply that for each $\varepsilon_1 > 0$ there exists a constant $\delta_0(\varepsilon_1, K_2)$ such that

$$(1.56) \quad 2\,\mathcal{I}m \int_{E_+^{n+1}} N\psi_1(0,D+i\eta)\psi u \, \overline{L(D+i\eta)\psi u} \, dx \geq$$

$$\geq 2\,\mathcal{I}m \int_{E_+^{n+1}} K(D+i\eta)\psi u \overline{L(D+i\eta)\psi u} \, dx - \varepsilon_1 |\eta| \, \|\psi u\|^2_{(1,0),\eta} -$$

$$- \varepsilon_1 \sum_{\mu=0}^{1} \| D_n^{\mu} \psi u(\cdot,0) \|^2_{(1-\mu),\eta} - K'' \|\psi u\|^2_{(1,0),\eta}$$

for all $u \in C^{\infty}_{(o)}(E_+^{n+1})$, $0 < \varepsilon < \delta_0(\varepsilon_1, K_2)$, $|\eta| > T_2(\varepsilon_0, \varepsilon, K_2)$.

The constants: K_1 in (1.39) and K_2 in (1.56) depend only on the parameter τ.

Finally consider the first term in the right hand side of the inequality (1.56). We have

$$(1.57)\qquad 2\,\Im m\int_{E_+^{n+1}} K(D+i\eta)\psi u\,\overline{L(D+i\eta)\psi u}\,dx =$$

$$= |\eta|\int_{E_+^{n+1}}\sum_{\mu,\nu=0}^{1} q_{\mu,\nu}(D'+i\eta')D_n^{\mu}\psi u\,\overline{D_n^{\nu}\psi u}\,dx +$$

$$+\int_{R_0^n}\sum_{\mu,\nu=0}^{1} p_{\mu,\nu}(D'+i\eta')D_n^{\mu}\psi u(x',0)\overline{D_n^{\nu}\psi u(x',0)}\,dx'$$

and $\sum_{\mu,\nu=0}^{1} q_{\mu,\nu}(\zeta')\zeta_n^{\mu}\overline{\zeta_n^{\nu}}$, $\sum_{\mu,\nu=0}^{1} p_{\mu,\nu}(\zeta')\zeta_n^{\mu}\overline{\zeta_n^{\nu}}$

are quadratic hermitian forms of the variables η_0 , ξ' and ζ_n. The first of these forms has a real coefficients, From the equalities (1.50) and (1.51) we get:

$$\sum_{\mu,\nu=0}^{1} q_{\mu,\nu}(\zeta)\xi_n^{\mu}\xi_n^{\nu} = \frac{2\,\Im m K(\zeta_0,\xi'',\xi_n)\overline{L(\zeta_0,\xi'',\xi_n)}}{|\eta_0|} =$$

$$= -a_0b_0\sum_{i=1}^{2}\mu_i(\xi'',\xi_n)\left|\zeta_0-\lambda_i(\xi'',\xi_n)\right|^2 \geq \gamma_0^{(\tau)}\left(|\zeta'|^2+\xi_n^2\right)$$

for some constant $\gamma_0^{(\tau)} > 0$ and an arbitrary real η , ξ . Thus the form $\sum_{\mu,\nu=0}^{1} q_{\mu,\nu}(\zeta')\zeta_n^{\mu}\overline{\zeta_n^{\nu}}$ is positive for real values of variables η_0 , ξ', ζ_n, and because it has the real coefficients it is positively definite also for any complex values of η_0 , ξ' and ζ_n. Hence

$$(1.58)\qquad \int_{E_+^{n+1}}\sum_{\mu,\nu=0}^{1} q_{\mu,\nu}(D'+i\eta')D_n^{\mu}\psi u\,\overline{D_n^{\nu}\psi u}\,dx \geq \gamma_0^{(\tau)}\Big(|\eta|^2\int_{E_+^{n+1}}|\psi u|^2\,dx +$$

$$+\sum_{j=0}^{n}\int_{E_+^{n+1}}|D_j\psi u|^2\,dx\Big) = \gamma_0^{(\tau)}\|\psi u\|^2_{(1,0),\eta}.$$

Combining (1.39), (1.56), (1.57), (1.58), estimating the term in the left hand side of (1.39) in a standard way, taking $\varepsilon_1 < \frac{1}{4}\gamma_0(\tau)$ and $|\eta|$ sufficiently large we get the required statement on the inequality (1.27). Thus it remains to prove (1.28). By (1.52) we have

$$2\,\mathrm{Im}\,K(\xi^{0\prime},\zeta_n)\overline{L(\xi^{0\prime},\zeta_n)} = 2\,\mathrm{Im}(\zeta_n-\xi_n^0)(\overline{\zeta_n-\xi_n^0+2\frac{\gamma_1}{\tau}}) =$$

$$= -i(\zeta_n-\bar{\zeta}_n)(|\zeta_n-\xi_n^0|^2+2\frac{\gamma_1}{\tau}(\zeta_n+\bar{\zeta}_n)-4\frac{\gamma_1}{\tau}\xi_n^0)$$

and it implies (1.28).

G.) Now let $\psi_2 \in C^\infty(Z^{n+1}\setminus\{0\})$ be a function homogeneous of degree 0 such that $(\mathrm{supp}\,\psi_2)\setminus\{0\} \subset C_0'$ and $\psi_2(\zeta')=1$ for $\zeta' \in (\mathrm{supp}\,\psi_1)\setminus\{0\}$.

In all the above considerations the functions ψ_2 and ψ_1 may be taken instead of ψ_1 and ψ respectively. Substitute the functions $R^+\psi_1(x,D+i\eta)\psi u$ and $R^-\psi_1(x,D+i\eta)\psi u$ for the function u in (1.18) and (1.19) respectively, similarly $R_j^+\psi_1(x,D+i\eta)\psi u$ in (1.23), $R_k^-\psi_1(x,D+i\eta)\psi u$ in (1.24) and $R_1\psi_1(x,D+i\eta)\psi u$ in (1.27). Using the proposition (i) and the definition of the polynomials B_j, and taking

$$\pi_{\mu,\nu}^{(l)}(\tau;\zeta') = |\zeta'|^{\mu+\nu-2}\,p_{\mu,\nu}^{(l)}(\tau;\zeta')$$

we obtain the following proposition :

for each $\varepsilon_1 > 0$ and $\tau > \tau_0$ there exists a constant $\delta_0(\varepsilon_1, \tau) > 0$ such that if $0 < \delta, \varepsilon < \delta_0(\varepsilon_1, \tau)$ then for some K

$$(1.59)\quad |\eta| \sum_{j=1}^{p} \int_{E_+^{n+1}} |B_j \psi_1(x, D+i\eta)\psi u|^2 \, dx \leq$$

$$\leq K\Big(\int_{E_+^{n+1}} |P^0(x, D+i\eta)\psi u|^2 \, dx + \sum_{j=1}^{p} \int_{R_0^n} |B_j \psi_1(x', 0, D+i\eta)\psi u|^2 dx' +$$

$$+ |||\psi u|||^2_{(m-2,0),\eta} \Big),$$

$$(1.60)\quad |\eta| \sum_{j=\varkappa+t+1}^{\varkappa+t+q} \int_{E_+^{n+1}} |B_j \psi_1(x, D+i\eta)\psi u|^2 \, dx +$$

$$+ \sum_{j=\varkappa+t+1}^{\varkappa+t+q} \int_{R_0^n} |B_j \psi_1(x', 0, D+i\eta)\psi u|^2 \, dx' \leq$$

$$\leq K\Big(\int_{E_+^{n+1}} |P^0(x, D+i\eta)\psi u|^2 \, dx + |||\psi u|||^2_{(m-2,0),\eta} \Big),$$

$$(1.61)\quad |\eta| \int_{E_+^{n+1}} |B_j \psi_1(x, D+i\eta)\psi u|^2 \, dx \leq$$

$$\leq K\Big(\int_{E_+^{n+1}} |P^0(x, D+i\eta)\psi u|^2 \, dx + \int_{R_0^n} |B_j \psi_1(x', 0, D+i\eta)\psi u|^2 \, dx' +$$

$$+ |||\psi u|||^2_{(m-2,0),\eta} \Big), \quad j = p+1, \ldots, p+r,$$

$$(1.62)\quad |\eta| \int_{E_+^{n+1}} |B_j \psi_1(x, D+i\eta)\psi u|^2 dx + \int_{R_0^n} |B_j \psi_1(x', 0, D+i\eta)\psi u|^2 dx' \leq$$

$$\leq K\Big(\int_{E_+^{n+1}} |P^0(x, D+i\eta)\psi u|^2 dx + |||\psi u|||^2_{(m-2,0),\eta} \Big),$$

$j = m-s+1,\ldots,m$,

$$(1.63)\quad \gamma_0(\tau)|\eta| \sum_{\mu=0}^{1} \int_{E_+^{n+1}} |B_{j+\mu t}\,\psi_1(x,D+i\eta)\psi u|^2\, dx \; +$$

$$+ \int_{R_0^n} \sum_{\mu,\nu=0}^{1} \pi_{\mu,\nu}^{(j-p-r)}(\tau,D'+i\eta')B_{j+\mu t}\,\psi_1(x',0,D+i\eta)\psi u\overline{B_{j+\nu t}\,\psi_1(x',0,D+i\eta)\psi u}\,dx' -$$

$$- \varepsilon_1 \sum_{\mu=0}^{1} \int_{R_0^n} |B_{j+\mu t}\,\psi_1(x',0,D+i\eta)\psi u|^2 dx' \leq$$

$$\leq K\left(\int_{E_+^{n+1}} |P^0(x,D+i\eta)\psi u|^2 dx \; + |||\psi u|||^2_{(m-2,0),\eta} \right),$$

$j = p+r+1,\ldots,p+r+t$,

for all $u \in C^{\infty}_{(0)}(E_+^{n+1})$, $|\eta| > T(\delta,\varepsilon,\varepsilon_1,\tau)$.

The constant K above depends on δ, ε, ε_1 and τ , but it is independent of u and η .

The form $\displaystyle\sum_{\mu,\nu=0}^{1} \pi_{\mu,\nu}^{(l)}(\tau;\zeta')z_\mu \overline{z_\nu}$, $(z_0,z_1) \in \mathbb{C}^2$, $l = 1,\ldots,t$, satisfies

$$(1.64)\quad \sum_{\mu,\nu=0}^{1} \pi_{\mu,\nu}^{(l)}(\tau;\xi^{0\prime})z_\mu \overline{z_\nu} = \left| z_1 - \frac{\xi^0_{n,l}}{|\xi^{0\prime}|} z_0 \right|^2 +$$

$$+ \frac{2}{\tau}\left(\frac{\gamma_1}{|\xi^{0\prime}|} z_1\overline{z_0} + \frac{\gamma_1}{|\xi^{0\prime}|} z_0\overline{z_1} - \frac{2\gamma_1 \xi^0_{n,l}}{|\xi^{0\prime}|} |z_0|^2 \right)$$

as it follows from (1.28).

Now recall the equality (1.15) and the definition of the matrix $C'(0,\xi^0)$ given at the end of the point C.). We have

$$(1.65)\quad \sum_{j=1}^{\varkappa} c'_{i,j}(0,\xi^{0\prime})Q_j(x',\zeta',\zeta_n) =$$

$$=\sum_{k=1}^{m}\left(\sum_{j=1}^{\varkappa} c'_{i,j}(0,\xi^{0\prime})b_{j,k}(x',\zeta')\right)B_k(x',\zeta',\zeta_n)$$

and, writing $c_{i,k}(x',\zeta')=\sum_{j=1}^{\varkappa} c'_{i,j}(0,\xi^{0\prime})b_{j,k}(x',\zeta')$,

$i = 1,\ldots,\varkappa$, $k = 1,\ldots,m$,

$$(1.66)\quad \sum_{j=1}^{\varkappa} c'_{i,j}(0,\xi^{0\prime})Q_j(x',D+i\eta)\psi u =$$

$$=\sum_{k=1}^{m} c_{i,k}\psi_1(x',D'+i\eta')B_k\psi_1(x',D+i\eta)\psi u + R_i\psi u \quad ,$$

$i = 1,\ldots,\varkappa$.

where $\|R_i\psi u\|^2_{(0),\eta}\le K'\sum_{l=0}^{m-2}\|D_n^l\psi u(\cdot,0)\|^2_{(m-2-l),\eta}$, and

$c_{i,k}\psi_1$ are pseudodifferential operators of order 0.

Introduce the denotations:

$$(1.67)\quad v_j(x) = B_j\psi_1(x,D+i\eta)\psi u \quad , \quad j = 1,\ldots,m ,$$

$$(1.68)\quad d_{j,k}(x',\zeta')=\sum_{i=1}^{\varkappa} c_{i,j}(x',\zeta')\overline{c_{i,k}(x',\zeta')} \quad ,$$

$j,k = 1,\ldots,m$.

Further we shall need the following result: because

$$\int_{R_0^n}\left|B_j\psi_1(x',0,D+i\eta)\psi_1\psi u - \psi_1 B_j\psi_1(x',0,D+i\eta)\psi u\right|^2 dx' \le$$

$$\leq K'\sum_{l=0}^{m-2}\|D_n^l \Psi u(\cdot,0)\|^2_{(m-2-l),\eta}$$

and $\Psi u = \Psi_1 \Psi u$, it is

$$(1.69)\quad \frac{1}{2}\int_{R_0^n} |\Psi_1 v_j(x',0)|^2 dx' - K'\sum_{l=0}^{m-2}\|D_n^l \Psi u(\cdot,0)\|^2_{(m-2-l),\eta} \leq$$

$$\leq \int_{R_0^n} |v_j(x',0)|^2 dx' \leq 2\int_{R_0^n} |\Psi_1 v_j(x',0)|^2 dx' +$$

$$+ K'\sum_{l=0}^{m-2}\|D_n^l \Psi u(\cdot,0)\|^2_{(m-2-l),\eta}, \quad j=1,\ldots,m .$$

By (1.66) and (1.67)

$$(1.70)\quad \sum_{i=1}^{\varkappa}\int_{R_0^n}\Big|\sum_{j=1}^{m} c_{i,j}\Psi_1(x',D'+i\eta')v_j\Big|^2 dx' \leq$$

$$\leq K\left(\sum_{j=1}^{\varkappa}\int_{R_0^n} |Q_j(x',D+i\eta)\Psi u|^2 dx' + |||\Psi u|||^2_{(m-2,0),\eta}\right).$$

Consider the left hand side of the above inequality. We have

$$(1.71)\quad \sum_{i=1}^{\varkappa}\int_{R_0^n}\Big|\sum_{j=1}^{m} c_{i,j}\Psi_2(x',D'+i\eta')\Psi_1 v_j\Big|^2 dx' =$$

$$= \sum_{j,k=1}^{m}\int_{R_0^n}\left(\sum_{i=1}^{\varkappa}(c_{i,k}\Psi_2)^*(x',D'+i\eta')c_{i,j}\Psi_2(x',D'+i\eta')\Psi_1 v_j \overline{\Psi_1 v_k}dx'\right) \geq$$

$$\geq \mathrm{Re}\int_{R_0^n}\sum_{j,k=1}^{m} d_{j,k}\Psi_2(x',D'+i\eta')\Psi_1 v_j\overline{\Psi_1 v_k}dx' -$$

$$- K'\sum_{j=1}^{m}\|\Psi_1 v_j(\cdot,0)\|^2_{(-\frac{1}{2}),\eta}$$

where it was used that $\psi_1 = \psi_2\psi_1 = \psi_2^2\psi_1$. The equalities (1.17), (1.65) and (1.68) imply that the quadratic form

$$\sum_{j,k=1}^{m} d_{j,k}(x',\zeta')z_j\overline{z_k}\ , \qquad z = (z_1,\dots,z_m) \in \mathbb{C}^m\ ,$$

is hermitian and satisfies

$$(1.72)\qquad \sum_{j,k=1}^{m} d_{j,k}(0,\xi^{0\prime})z_j\overline{z_k} = \sum_{i=1}^{\varkappa}\left|\sum_{j=1}^{m} c_{i,j}(0,\xi^{0\prime})z_j\right|^2 =$$

$$= \sum_{i=1}^{\varkappa}\left| z_i + \sum_{l=1}^{t} c_{i,\varkappa+1}(0,\xi^{0\prime})(z_{\varkappa+1} - \frac{\xi^0_{n,l}}{|\xi^{0\prime}|}\, z_{\varkappa-t+1}) + \right.$$

$$\left. + \sum_{k=\varkappa+t+1}^{m} c_{i,k}(0,\xi^{0\prime})z_k\right|^2 .$$

Combining (1.71) and (1.72) we get:

$$(1.73)\qquad \mathrm{Re}\int_{R^n_0}\sum_{j,k=1}^{m} d_{j,k}\psi_2(x',D'+i\eta')\psi_1 v_j\overline{\psi_1 v_k}\,dx' -$$

$$- \frac{K'}{|\eta|}\sum_{j=1}^{m}\int_{R^n_0}\left|\psi_1 v_j(x',0)\right|^2 dx' \le$$

$$\le K\left(\sum_{j=1}^{\varkappa}\int_{R^n_0}\left|Q_j(x',D+i\eta)\psi u\right|^2 dx' + |||\overset{(\psi)}{u}|||^2_{(m-2,0),\eta}\right).$$

Now multiply (1.59) and (1.61) by $\frac{\varepsilon_1}{K'}$, (1.73) by $\frac{1}{\lambda}$, $\lambda > 1$, and sum so obtained inequalities with (1.60), (1.62) and (1.63). Remove the all boundary integrals in the left hand side of this inequality. Thus, assuming that $\frac{\varepsilon_1}{K} < 1$ and $\gamma_0(\tau) < 1$, using (1.67) and (1.69), we have the following statement:

for each $\varepsilon_1 > 0$ and $\tau > \tau_0$ there exists a constant $\delta_0(\varepsilon_1, \tau) > 0$ such that if $0 < \delta, \varepsilon < \delta_0(\varepsilon_1, \tau)$, then for some K

$$(1.74)\quad \frac{\varepsilon_1 \gamma_0(\tau)}{K} |\eta| \sum_{j=1}^{m} \int_{E_+^{n+1}} |v_j(x)|^2 dx + $$

$$+ \operatorname{Re} \int_{R_0^n} \sum_{j,k=1}^{m} p_{j,k} \psi_2(x', D' + i\eta') \psi_1 v_j \overline{\psi_1 v_k}\, dx' -$$

$$- \frac{K'}{\lambda |\eta|} \sum_{j=1}^{m} \int_{R_0^n} |\psi_1 v_j(x', 0)|^2 dx' \leq$$

$$\leq K \left(\int_{E_+^{n+1}} |P^0(x, D + i\eta) \psi u|^2 dx + \sum_{j=1}^{\varkappa} \int_{R_0^n} |Q_j(x, D' + i\eta') \psi u|^2 dx' + \right.$$

$$\left. + |||\psi u|||^2_{(m-2,0),\eta} \right)$$

holds for all $u \in C^{\infty}_{(0)}(E_+^{n+1})$, $|\eta| > T(\delta, \varepsilon, \varepsilon_1, \tau)$, $\lambda > 1$;

and

$$(1.75)\quad \sum_{j,k=1}^{m} p_{j,k}(x', \zeta') z_j \overline{z_k} = \sum_{j=\varkappa+t+1}^{m} |z_j|^2 +$$

$$+ \sum_{j=p+r+1}^{p+r+t} \sum_{\mu,\nu=0}^{1} \pi_{\mu,\nu}^{(j-p-r)}(\tau; \zeta') z_{j+\mu t} \overline{z_{j+\nu t}} +$$

$$+ \frac{1}{\lambda} \sum_{j,k=1}^{m} d_{j,k}(x', \zeta') z_j \overline{z_k} - 3\varepsilon_1 \sum_{j=1}^{m} |z_j|^2 ,$$

$z = (z_1, \dots, z_m) \in \mathbb{C}^m$,

is a quadratic hermitian form with smooth coefficients, homogeneous of degree 0 in ζ'.

We shall show that for suitably chosen values of the parameters $\varepsilon_1, \tau, \delta, \varepsilon$ and λ this quadratic form is positively definite, e.g. for some $\gamma_0 > 0$

$$(1.76) \qquad \sum_{j,k=1}^{m} p_{j,k}(x', \zeta') z_j \overline{z_k} \geqslant 2\gamma_0 \sum_{j=1}^{m} |z_j|^2 ,$$

$z \in \mathbb{C}^m$, $(x', \zeta') \in R_0^n \times C_0'$.

By the homogeneity and continuity of the coefficients $p_{j,k}$ it is sufficient to prove it for the point $(0, \xi^{0\prime})$, provided that δ and ε are sufficiently small.

Since (1.64) , (1.72) and (1.75) we have:

$$\sum_{j,k=1}^{m} p_{j,k}(0, \xi^{0\prime}) z_j \overline{z_k} = \sum_{j=\varkappa+t+1}^{m} |z_j|^2 +$$

$$+ \sum_{j=p+r+1}^{p+r+t} \Big(\Big| z_{j+t} - \frac{\xi^0_{n, j-p-r}}{|\xi^{0\prime}|} z_j \Big|^2 +$$

$$+ \frac{2}{\tau} \Big(\frac{\gamma_1}{|\xi^{0\prime}|} z_{j+t} \overline{z_j} + \frac{\gamma_1}{|\xi^{0\prime}|} z_j \overline{z_{j+t}} - \frac{2\gamma_1 \xi^0_{n, j-p-r}}{|\xi^{0\prime}|} |z_j|^2 \Big) \Big) +$$

$$+ \frac{1}{\lambda} \sum_{i=1}^{\varkappa} \Big| z_i + \sum_{l=1}^{t} c_{i, \varkappa+l}(0, \xi^{0\prime}) \Big(z_{\varkappa+l} - \frac{\xi^0_{n,l}}{|\xi^{0\prime}|} z_{\varkappa-t+l} \Big) +$$

$$+ \sum_{k=\varkappa+t+1}^{m} c_{i,k}(0, \xi^{0\prime}) z_k \Big|^2 - 3\varepsilon_1 \sum_{j=1}^{m} |z_j|^2 \geqslant \sum_{j=\varkappa+t+1}^{m} |z_j|^2 +$$

$$+ \sum_{l=1}^{t} \left| z_{\varkappa+l} - \frac{\xi^0_{n,l}}{|\xi^{0'}|} z_{\varkappa-t+l} \right|^2 - \frac{c_0}{\tau} \sum_{j=\varkappa-t+1}^{\varkappa+t} |z_j|^2 +$$

$$+ \frac{1}{2\lambda} \sum_{i=1}^{\varkappa} |z_i|^2 - \frac{c_0}{\lambda} \sum_{l=1}^{t} \left| z_{\varkappa+l} - \frac{\xi^0_{n,l}}{|\xi^{0'}|} z_{\varkappa-t+l} \right|^2 -$$

$$- \frac{c_0}{\lambda} \sum_{j=\varkappa+t+1}^{m} |z_j|^2 - 3\varepsilon_1 \sum_{j=1}^{m} |z_j|^2 .$$

Fix a number $\lambda > 1$ in such a way, that $1 - \frac{c_0}{\lambda} > \frac{1}{2}$.

Then

$$\sum_{j,k=1}^{m} p_{j,k}(0,\xi^{0'}) z_j \overline{z_k} \geqslant \frac{1}{2\lambda} \sum_{j=1}^{\varkappa-t} |z_j|^2 + \frac{1}{2} \sum_{j=\varkappa+t+1}^{m} |z_j|^2 +$$

$$+ \sum_{l=1}^{t} \left(\frac{1}{2} \left| z_{\varkappa+l} - \frac{\xi^0_{n,l}}{|\xi^{0'}|} z_{\varkappa-t+l} \right|^2 + \frac{1}{2\lambda} |z_{\varkappa-t+l}|^2 \right) -$$

$$- \frac{c_0}{\tau} \sum_{j=\varkappa-t+1}^{\varkappa+t} |z_j|^2 - 3\varepsilon_1 \sum_{j=1}^{m} |z_j|^2 .$$

The quadratic form $\frac{1}{2} \left| z_{\varkappa+l} - \frac{\xi^0_{n,l}}{|\xi^{0'}|} z_{\varkappa-t+l} \right|^2 + \frac{1}{2\lambda} |z_{\varkappa-t+l}|^2$

is positively definite, therefore for some $\gamma_0 > 0$

$$\sum_{j,k=1}^{m} p_{j,k}(0,\xi^{0'}) z_j \overline{z_k} \geqslant 4\gamma_0 \sum_{j=1}^{m} |z_j|^2 - \frac{c_0}{\tau} \sum_{j=\varkappa-t+1}^{\varkappa+t} |z_j|^2 - 3\varepsilon_1 \sum_{j=1}^{m} |z_j|^2 ,$$

and choosing $\tau > \tau_0$ and $\varepsilon_1 > 0$ such that $\gamma_0 - \frac{c_0}{\tau} - 3\varepsilon_1 \geqslant 0$

we have

$$\sum_{j,k=1}^{m} p_{j,k}(0,\xi^{0'}) \geqslant 3\gamma_0 \sum_{j=1}^{m} |z_j|^2 .$$

Thus we have fixed the numbers λ, τ and ε_1 and now δ, ε can be fixed so that all the above statements are valid.

The proposition (iii), (1.76) and (1.69) give

$$(1.77)\qquad \operatorname{Re}\int_{R_0^n} \sum_{j,k=1}^{m} p_{j,k}\Psi_2(x', D'+i\eta')\psi_1 v_j \overline{\psi_1 v_k}\, dx' -$$

$$- \frac{K'}{\lambda|\eta|}\sum_{j=1}^{m}\int_{R_0^n} \left|\psi_1 v_j(x',0)\right|^2 dx' \geqslant$$

$$\geqslant \frac{1}{2}\gamma_0 \sum_{j=1}^{m}\int_{R_0^n}\left|v_j(x',0)\right|^2 dx' - K\,|||\psi u|||^2_{(m-2,0),\eta}$$

for $|\eta|$ large.

Finally, (1.14) and the proposition (i) give

$$(1.78)\qquad \|\psi u\|^2_{(m-1,0),\eta} \leq K\left(\sum_{j=1}^{m}\int_{E_+^{n+1}} |v_j(x)|^2 dx + \|\psi u\|^2_{(m-2,0),\eta}\right),$$

$$(1.79)\qquad \sum_{l=0}^{m-1}\left\| D_n^{l}\psi u(\cdot,0)\right\|^2_{(m-1-l),\eta} \leqslant$$

$$\leqslant K\left(\sum_{j=1}^{m}\int_{R_0^n}\left|v_j(x',0)\right|^2 dx' + \sum_{l=0}^{m-2}\left\| D_n^{l}\psi u(\cdot,0)\right\|^2_{(m-2-l),\eta}\right)$$

From the inequalities (1.74), (1.77), (1.78) and (1.79) we get

$$|||\psi u|||^2_{(m-1,0),\eta} \leq K\Bigl(\int_{E_+^{n+1}} |P^o(x,D+i\eta)\psi u|^2 \, dx \; +$$

$$+ \sum_{j=1}^{\varkappa} \int_{R_o^n} |Q_j(x',D+i\eta)\psi u|^2 \, dx' \; +$$

$$+ |||\psi u|||^2_{(m-2,0),\eta} \Bigr) \quad .$$

Because $|||\psi u|||^2_{(m-2,0),\eta} \leq \frac{1}{|\eta|^2} |||\psi u|||^2_{(m-1,0),\eta}$, it is

(1.80) $$|||\psi u|||^2_{(m-1,0),\eta} \leq K\Bigl(\|P^o\psi u\|^2_{(o,o),\eta} + \sum_{j=1}^{\varkappa} \|Q_j\psi u\|^2_{(o),\eta}\Bigr)$$

for all $u \in C^\infty_{(o)}$ and $|\eta|$ sufficiently large.

Let us consider the family of all open cones $C' \subset Z^{n+1}$ such, that the inequality (1.80) holds for each function $\psi \in C^\infty(Z^{n+1} \setminus \{0\})$ homogeneous of degree 0 and with $(\mathrm{supp}\,\psi) \setminus \{0\} \subset C'$.

This family covers the set $\overline{Z_-^{n+1}} \setminus \{0\}$. To prove it, it must be shown that for each point $\zeta' \in \overline{Z_-^{n+1}} \setminus \{0\}$ there exists a cone C' from this family which is the neighbourhood of ζ'. For the points ζ' with $\eta_o = 0$ and $\xi'' \neq 0$ it was proved above. When $\eta_o \neq 0$, then this statement follows from the remark at the end of the point D.) and from all the later considerations. Thus it remains to consider the points $(1,0,\ldots,0)$ and $(-1,0,\ldots,0)$. In some neighbourhoods of these points the polynomial $P^o(x, \zeta', \zeta_n)$ has the factorisation (1.10) with the polynomials $M_j^+(x, \zeta', \zeta_n)$ and $M_k^-(x, \zeta', \zeta_n)$ only occuring in it. Then the results of the point E (and the considerations of G) in a simplified form show the existence of the desired cones for these points.

Choose a finite subfamily which covers the set $\overline{Z_-^{n+1}} \setminus \{0\}$ and let functions Ψ_j, $j=1,\dots,N$, are smooth, homogeneous of degree 0, and form a partition of unity on $\overline{Z_-^{n+1}} \setminus \{0\}$ subordinated to the chosen subfamily of cones.

Then we can fix $\delta > 0$ common for all functions Ψ_j and the standard argument with a partition of unity shows that the inequality (1.8) holds.

Thus the Theorem 1 is proved.

We shall need a slightly generalized version of Theorem 1. Denote $m_o = \max_{1 \le j \le \varkappa} m_j$, $\mathcal{H}_{(\tau,-\infty)}(E_+^{n+1}) = \bigcup_s \mathcal{H}_{(\tau,s)}(E_+^{n+1})$, and $\mathcal{B}_{(k,s)}(E_+^{n+1})$, k is an nonnegative integer, denotes the space of all $u \in \mathcal{H}_{(k+1,s-1)}(E_+^{n+1})$ such that $D_n^l u(\cdot,0) \in \mathcal{H}_{(k-l+s)}(R_o^n)$ for $l = 0,\dots,k$, with the norm $|||u|||_{(k,s),\eta}$.

Theorem 2. If the assumptions of Theorem 1, except that now m_j are arbitrary nonnegative integers, are satisfied for the operators $P(x, D+i\eta)$ and $Q_j(x', D+i\eta)$, $j=1,\dots,\varkappa$, k is a nonnegative integer such that $m-1+k \geqslant m_0$, and s is arbitrary real number, then the inequality

$$|||u|||_{(m-1+k,s),\eta} \leq K\left(\|P_\eta u\|^2_{(k.s),\eta} + \sum_{j=1}^{\varkappa} \| Q_{j,\eta}u(\cdot,0)\|^2_{(m-1+k-m_j+s),\eta}\right)^{\frac{1}{2}}$$

holds for all $u \in \mathcal{H}_{(m+k,-\infty)}(E^{n+1}_+)$ and for all sufficiently large $|\eta|$, $\eta_0 < 0$, with some constant K independent of u and η .

Remark. The both sides of (1.81) have a sense for any $u \in \mathcal{H}_{(m+k,-\infty)}(E^{n+1}_+)$ and, of course, can be equal to $+\infty$. Theorem is interesting only for u such that right hand side of (1.81) is finite.

Proof. At first we shall prove (1.81) for $u \in C^\infty_{(0)}(E^{n+1}_+)$. We can write

$$(1.82) \quad Q_j(x',\zeta',\zeta_n) = A_j(x',\zeta',\zeta_n)P^0(x',\zeta',\zeta_n) + R_j(x',\zeta',\zeta_n),$$

$j = 1,\ldots,\varkappa$,

where $A_j(x',\zeta',\zeta_n)$, $R_j(x',\zeta',\zeta_n)$ are polynomials in ζ_n, which coefficients are symbols of pseudodifferential operators. They are of degree m_j-m and m_j respectively, but the degree of $R_j(x',\zeta',\zeta_n)$ with respect to ζ_n is $< m$. Defining

$$(1.83) \quad Q'_j(x',\zeta',\zeta_n) = R_j(x',\zeta',\zeta_n)|\zeta'|^{m-1-m_j} , \quad j=1,\ldots,\varkappa,$$

we have the operators $Q_j'(x',D+i\eta)$ which satisfy all the assumptions of Theorem 1, thus the inequality (1.5) holds with $Q_j'(x',D+i\eta)$ instead of $Q_j(x',D+i\eta)$. Substituting in it the function $\Lambda_{(k+s),\eta}u$ for the function u and estimating the commutators by the proposition (i) we obtain

$$|||u|||^2_{(m-1,k+s),\eta} \leq K\left(\|P_\eta u\|^2_{(0,k+s),\eta} + \sum_{j=1}^{\varkappa}\|Q'_{j,\eta}\, u(\cdot,0)\|^2_{(k+s),\eta}\right).$$

From Theorem 4.3.1 of [4] it follows that the inequality (1.81) holds for the operators $Q'_{j,\eta}$ and $u \in C^\infty_{(0)}(E^{n+1}_+)$. Now, using (1.82) and (1.83), the right hand side of this inequality can be estimated by a constant multiplied by

$$\|P_\eta u\|^2_{(k,s),\eta} + \sum_{j=1}^{\varkappa}\|Q_{j,\eta}\, u(\cdot,0)\|^2_{(m-1+k-m_j+s),\eta} + |||u|||^2_{(m-2+k,s),\eta},$$

hence (1.81) holds for $Q_{j,\eta}$ also.

Let $\Lambda_{\eta,\delta}$ be the operator given by the symbol $(1+\delta^2|\zeta'|^2)^{-\frac{1}{2}}$. We shall need the following result.

Lemma. If $p(x',D'+i\eta)$ is a pseudodifferential operator of order r, then for any real number t there exists a constant C such that

$$(1.84)\qquad \|p\Lambda_{\eta,\delta}u - \Lambda_{\eta,\delta}pu\|_{(t),\eta} \leq C\|\Lambda_{\eta,\delta}u\|_{(r+t-1),\eta}$$

for all $u \in \mathcal{H}_{(r+t-1)}(R^n)$, $|\eta| \geq 1$, $0 < \delta < 1$, and

(1.85) $$\|p\Lambda_{\eta,\delta}u - \Lambda_{\eta,\delta}pu\|_{(t),\eta} \leq C\delta\|\Lambda_{\eta,\delta}u\|_{(\tau+t),\eta}$$

for all $u \in \mathcal{H}_{(\tau+t)}(R^n)$, $|\eta| \geq 1$, $0 < \delta < 1$.

Proof. Let $p(x', \zeta') = p(\infty, \zeta') + p'(x', \zeta')$ and $p'(\cdot, \zeta') \in C_0^\infty(R^n)$ for $\zeta' \neq 0$, moreover supp $p'(\cdot, \zeta')$ is contained in a fixed compact subset of R^n. Then we have:

$$\|p\Lambda_{\eta,\delta}u - \Lambda_{\eta,\delta}pu\|^2_{(t),\eta} = (2\pi)^{-3n}\int |\zeta'|^{2t}\Big|\int \hat{p}'(\xi' - \Theta', \Theta' + i\eta')\cdot$$

$$\cdot \left(1 + \delta^2(\eta_0^2 + |\Theta'|^2)\right)^{-\frac{1}{2}} \hat{u}(\Theta')d\Theta' -$$

$$- \left(1 + \delta^2(\eta_0^2 + |\xi'|^2)\right)^{-\frac{1}{2}} \int \hat{p}'(\xi' - \Theta', \Theta' + i\eta')\hat{u}(\Theta')d\Theta'\Big|^2 d\xi' =$$

$$= (2\pi)^{-3n}\int |\zeta'|^{2t}\Big|\int \hat{p}'(\xi' - \Theta', \Theta' + i\eta')\int_0^1 \frac{\delta^2 \sum_{i=0}^{n-1}\left(\Theta_i + \tau(\xi_i - \Theta_i)\right)(\xi_i - \Theta_i)}{\left(1 + \delta^2(\eta_0^2 + |\Theta' + \tau(\xi' - \Theta')|^2)\right)^{\frac{3}{2}}} d\tau\, \hat{u}(\Theta')d\Theta'\Big|^2 d\xi' \leq$$

$$\leq C_p\int |\zeta'|^{2t}\Big|\int (1 + |\xi' - \Theta'|)^{-p+1}(\eta_0^2 + |\Theta'|^2)^{\frac{1}{2}}\int_0^1 \frac{\delta\, d\tau}{1 + \delta^2(\eta_0^2 + |\Theta' + \tau(\xi' - \Theta')|^2)}|\hat{u}(\Theta')|d\Theta'\Big|^2 d\xi'.$$

For any vectors $\xi', \Theta \in R^n$ and $|\eta| \geq 1$ we have

$$1 + \delta^2(\eta_0^{\,2} + |\xi' + \theta'|^2) \leq 2(1 + \delta^2(\eta_0^{\,2} + |\xi'|^2 + |\theta'|^2)) \leq$$

$$\leq 2(1 + \delta^2(\eta_0^{\,2} + |\xi'|^2))(1 + |\theta'|)^2$$

hence

$$(1 + \delta^2(\eta_0^{\,2} + |\xi' + \theta'|^2))^{-1} \leq 2(1 + \delta^2(\eta_0^2 + |\xi'|^2))^{-1}(1 + |\theta'|)^2 ,$$

similarly

$$|\zeta'|^{2t} \leq 2^{|t|}(\eta_0^2 + |\theta'|^2)^t (1 + |\xi' - \theta'|)^{2|t|} ,$$

and applying these estimations to the above inequality we obtain

$$\|p\Lambda_{\eta,\delta}u - \Lambda_{\eta,\delta}pu\|^2_{(t),\eta} \leq C'_p \int \left| \int (1 + |\xi' - \theta'|)^{-p+3+|t|} \frac{\delta(\eta_0^2 + |\theta'|^2)^{\frac{r+t}{2}}}{1 + \delta^2(\eta_0^2 + |\theta'|^2)} |\hat{u}(\theta')| d\theta' \right|^2 d\xi' .$$

Now (1.84) and (1.85) can be easily obtained by estimating

$$\frac{\delta(\eta_0^2 + |\theta'|^2)^{\frac{r+t}{2}}}{1 + \delta^2(\eta_0^2 + |\theta'|^2)} \text{ by } (1 + \delta^2(\eta_0^2 + |\theta'|^2))^{-\frac{1}{2}} (\eta_0^2 + |\theta'|^2)^{\frac{r+t-1}{2}} ,$$

respectively by $\delta(\eta_0^2 + |\theta'|^2)^{\frac{r+t}{2}}$ and using the standard integral inequality (generalized Minkowski's inequality). The proof of lemma is completed.

From (1.85) we conclude that if $u \in \mathcal{H}_{(r+t)}(R^n)$ and $\delta \longrightarrow 0^+$, then $p\Lambda_{\eta,\delta}u - \Lambda_{\eta,\delta}\, pu \longrightarrow 0$ in the topology of of the space $\mathcal{H}_{(t)}(R^n)$. Because (1.84) implies that the operators $p\Lambda_{\eta,\delta} - \Lambda_{\eta,\delta}p$, considered as the operators from the space $\mathcal{H}_{(r+t-1)}(R^n)$ into $\mathcal{H}_{(t)}(R^n)$, are uniformly bounded for $0 < \delta < 1$, we have $p\Lambda_{\eta,\delta}u - \Lambda_{\eta,\delta}pu \longrightarrow 0$ in $\mathcal{H}_{(t)}(R^n)$ for all $u \in \mathcal{H}_{(r+t-1)}(R^n)$.

Applying (1.84), (1.85) and the above conclusion to the operators P_η and $Q_{j,\eta}$ we obtain

(1.86) $$\| P_\eta\Lambda_{\eta,\delta}u - \Lambda_{\eta,\delta}P_\eta u\|^2_{(k,t),\eta} \leq c\|\Lambda_{\eta,\delta}u\|^2_{(m-1+k,t),\eta}$$

and $P_\eta\Lambda_{\eta,\delta}u - \Lambda_{\eta,\delta}P_\eta u \longrightarrow 0$ in $\mathcal{H}_{(k,t)}(E^{n+1}_+)$ for all $u \in \mathcal{H}_{(m-1+k,t)}(E^{n+1}_+)$, similarly

(1.87) $$\| Q_{j,\eta}\Lambda_{\eta,\delta}u(\cdot,0) - \Lambda_{\eta,\delta}Q_{j,\eta}u(\cdot,0)\|^2_{(m-1+k-m_j+t),\eta} \leq$$

$$\leq c\sum_{l=0}^{m-1+k}\|\Lambda_{\eta,\delta}D_n^l u(\cdot,0)\|^2_{(m-1+k-l+t-1),\eta}$$

and $Q_{j,\eta}\Lambda_{\eta,\delta}u(\cdot,0) - \Lambda_{\eta,\delta}Q_{j,\eta}u(\cdot,0) \longrightarrow 0$ in $\mathcal{H}_{(m-1+k-m_j+t)}(R^n_o)$ for all $u \in \mathcal{B}_{(m-1+k,t-1)}(E^{n+1}_+)$.

The next step is to prove the following statement: for any real number t there is a constant K such that

$$(1.88)\quad |||\Lambda_{\eta,\delta}u|||^2_{(m-1+k,t+1),\eta} \leq K\Big(\|\Lambda_{\eta,\delta}P_\eta u\|^2_{(k,t+1),\eta} + \sum_{j=1}^{\varkappa}\|\Lambda_{\eta,\delta}Q_{j,\eta}u(\cdot,0)\|^2_{(m-1+k-m_j+t+1),\eta}\Big)$$

for all δ , $0 < \delta < 1$, $|\eta|$ sufficiently large, and for all $u \in \mathcal{B}_{(m-1+k,t)}(E_+^{n+1})$. Substituting $\Lambda_{\eta,\delta}u$ for u in (1.81) with $t+1$ instead of s and estimating the commutators by (1.86) and (1.87) we obtain (1.88) with additional term $\|\Lambda_{\eta,\delta}u\|^2_{(m-1+k,t+1),\eta} + \sum_{l=0}^{m-1+k}\|\Lambda_{\eta,\delta}D_n^l u(\cdot,0)\|^2_{(m-1+k-l+t),\eta}$ on the right hand side. Because this term can be estimated by $\frac{1}{|\eta|}|||\Lambda_{\eta,\delta}u|||^2_{(m-1+k,t+1),\eta}$ (1.88) holds for $|\eta|$ sufficiently large and $u \in C^\infty_{(o)}(E_+^{n+1})$. The norms used in (1.88) are continuous in the topology of the space $\mathcal{H}_{(m+k,t)}(E_+^{n+1})$, hence it holds also for u from this space. Finally, let $u \in \mathcal{B}_{(m-1+k,t)}(E_+^{n+1})$, then $\Lambda_{\eta,\gamma}u \in \mathcal{H}_{(m+k,t)}(E_+^{n+1})$, $0 < \gamma < 1$, and we can apply (1.88) to this function. Taking $\gamma \to 0^+$ and using the conclusions following (1.86) and (1.87) we obtain the required statement.

Now we are ready to prove Theorem 2 in whole generality. Let $u \in \mathcal{H}_{(m+k,-\infty)}(E_+^{n+1})$ be such that $P_\eta u \in \mathcal{H}_{(k,s)}(E_+^{n+1})$

and $Q_{j,\gamma}u(\cdot,0) \in \mathcal{H}_{(m-1+k-m_j+s)}(R^n_o)$. Then for some number s_1 $u \in \mathcal{B}_{(m-1+k,s_1)}(E^{n+1}_+)$ and $s-s_1$ is a nonnegative integer. It is easily seen that it suffices to prove the following statement:

if for some t $u \in \mathcal{B}_{(m-1+k,t)}(E^{n+1}_+)$ and $t+1 \leq s$, then $u \in \mathcal{B}_{(m-1+k,t+1)}(E^{n+1}_+)$ and (1.81) holds with $t+1$ instead of s. This statement follows immediately from (1.88), if we put $\delta \longrightarrow 0^+$ in it and next apply Theorem 4.3.1 of [4].

2. The dual inequality.

Let H denotes the space $\mathcal{H}_{(k,s)}(E^{n+1}_+) \oplus \bigoplus_{j=1}^{\varkappa} \mathcal{H}_{(m-1+k-m_j+s)}(R^n_o)$ with the norm $\|F\|_{H,\gamma} = \left(\|f\|^2_{(k,s),\gamma} + \sum_{j=1}^{\varkappa} \|g_j\|^2_{(m-1+k-m_j+s),\gamma}\right)^{\frac{1}{2}}$ for $F = (f,g_1,\dots,g_{\varkappa})$. Consider the operator T

$$T : \mathcal{D}_T \ni u \longrightarrow (P_\gamma u, Q_{1,\gamma}u(\cdot,0),\dots,Q_{\varkappa,\gamma}\, u(\cdot,0)) \in H .$$

The domain $\mathcal{D}_T$ is the set of all $u \in \mathcal{B}_{(m-1+k,s)}(E^{n+1}_+)$ such that $P_\gamma u \in \mathcal{H}_{(k,s)}(E^{n+1}_+)$. Now the inequality (1.81) may be written in the form

(2.1) $$|||u|||_{(m-1+k,s),\gamma} \leq K \|Tu\|_{H,\gamma}, \quad u \in \mathcal{D}_T ,$$

furthermore

$$(2.2) \qquad \mathcal{H}_{(m+k,s)}(E_+^{n+1}) \subset \mathcal{D}_T \subset \mathcal{H}_{(m-1+k,s)}(E_+^{n+1}) .$$

Combining (2.1), (2.2) and Theorem 2 we get: T is a closed linear operator from $\mathcal{H}_{(m-1+k,s)}(E_+^{n+1})$ to H with a dense domain $\mathcal{D}_T$, and the operator T^{-1} is well defined and continuous. Hence $T\mathcal{D}_T$ is a closed linear subspace of H. Our aim is to prove that $T\mathcal{D}_T$ is equal to H. This is equivalent with the fact that the adjoint operator T^* is invertible, or what is the same T^* fulfils an inequality similar to (2.1).

It follows from general theory that T^* is closed, defined on a dense subset $\mathcal{D}_{T^*} \subset H^*$, and values of T^* are in $\mathcal{H}^*_{(m-1+k,s)}(E_+^{n+1})$. The space H^* is isomorphic with $\mathring{\mathcal{H}}_{(-k,-s)}(\overline{E_+^{n+1}}) \oplus \bigoplus_{j=1}^{\varkappa} \mathcal{H}_{(-m+1-k+m_j-s)}(R_o^n)$ by the duality given by the extension of the bilinear form:

$$(2.3) \qquad \int_{E_+^{n+1}} f(x)\overline{v(x)}dx + \sum_{j=1}^{\varkappa} \int_{R_o^n} g_j(x')\overline{v_j(x')}dx' ,$$

$$f \in C^\infty_{(o)}(E_+^{n+1}) , \quad v \in C_o^\infty(E_+^{n+1}) , \quad g_j, v_j \in C_o^\infty(R_o^n) .$$

We shall identify these two spaces.

Let us recall that if $V = (v, v_1, \dots, v_\varkappa) \in H^*$, then $T^* V$

is the linear functional on $\mathcal{D}_T$ defined by the extension of (2.3):

$$(2.4)\qquad (T^* V)(u) = \langle P_\eta u, v\rangle + \sum_{j=1}^{\varkappa} \langle Q_{j,\eta} u(\cdot,0), v_j\rangle ,$$

$$u \in C_{(0)}^{\infty}(E_+^{n+1}) ,$$

and the domain $\mathcal{D}_{T^*}$ is the set of all $V \in H^*$ such that this functional is continuous in the topology of $\mathcal{H}_{(m-1+k,s)}(E_+^{n+1})$.

At first we shall consider the case $m_j < m$, $j = 1,\dots,\varkappa$, and $k = 0$.

It is possible to prove an analogous inequality to (2.1) for T^*, but it is simpler to do it for this operator restricted to a dense linear subset of H^* which T^* maps into sufficiently regular functionals. More exactly, let $\mathcal{D}_0$ denote the set of all $V \in H^*$ such that $T^* V$ is a continuous linear functional in the norm

$$\|u\|_0 = \left(\|u\|^2_{(0,s),\eta} + \sum_{l=0}^{m-1} \|D_n^l u(\cdot,0)\|^2_{(-l+s),\eta}\right)^{\frac{1}{2}},$$

$u \in C_{(0)}^{\infty}(E_+^{n+1})$.

Theorem 3. Suppose that $P(x, D-i\eta)$ and $Q_j(x', D+i\eta)$, $j = 1,\dots,\varkappa$, verify the assumptions of Theorem 1, and let s be any real number. Then there exists a constant K such that for all $v \in \mathcal{H}_{(o,-s)}(E_+^{n+1})$, $v_j \in \mathcal{H}_{(-m+1+m_j-s)}(R_o^n)$ and $|\eta|$ sufficiently large there holds inequality

$$(2.5) \quad \left(|||v|||^2_{(m-1,-s),\eta} + \sum_{j=1}^{\varkappa} \|v_j\|^2_{(m_j-s),\eta} \right)^{\frac{1}{2}} \leq$$

$$\leq K \sup_{u \in C^\infty_{(o)}(E_+^{n+1})} \frac{|(T^* V)(u)|}{\|u\|_o} .$$

Remark. Of course the inequality (2.5) is interesting only for $V \in \mathcal{D}_o$, for others V it is trivially verified because the right hand side of (2.5) is then equal to $+\infty$.

Proof. Let

$$P(x, D+i\eta) = \sum_{l=0}^{m} p_l(x, D' + i\eta') D_n^{m-l} ,$$

$$Q_j(x', D+i\eta) = \sum_{l=0}^{m_j} q_{j,l}(x', D'+i\eta') D_n^{m-1-l} ,$$

$p_l(x, D'+i\eta')$ and $q_{j,l}(x', D'+i\eta')$ be differential and pseudo-differential operators of order l and $m_j+l-m+1$ respectively, such that if $m_j+l-m+1 < 0$, then $q_{j,l}(x', D'+i\eta')$ is a zero operator. Let us put

$$P_l(x', D+i\eta) = \sum_{k=0}^{l} p_k(x', D'+i\eta') D_n^{l-k} , \quad l = 0, 1, \ldots, m-1 .$$

For $v \in C^{\infty}_{(o)}(E^{n+1}_{+})$ and $v_j \in C^{\infty}_{o}(R^{n}_{o})$ we have :

$$\begin{aligned}(T^{*} V)(u) &= \int_{E^{n+1}_{+}} P(x,D+i\eta)u(x)\overline{v(x)}dx + \\ &+ \sum_{j=1}^{\varkappa} \int_{R^n_o} Q_j(x',D+i\eta)u(x',0)\overline{v_j(x')}dx' = \\ &= \int_{E^{n+1}_{+}} u(x)\, \overline{P^{*}(x,D+i\eta)v(x)}dx + \\ &+ \sum_{l=0}^{m-1} \int_{R^n_o} D_n^{m-1-l}u(x',0)\overline{\Big(i\sum_{k=0}^{l} D_n^{l-k}p_k^{*}(x',D'+i\eta')v(x',0) +} \\ &\overline{+ \sum_{j=1}^{\varkappa} q^{*}_{j,l}(x',D'+i\eta')v_j(x')\Big)}dx', \qquad u \in C^{\infty}_{(o)}(E^{n+1}_{+}) .\end{aligned}$$

This equality can be extended for any $v \in \mathcal{H}_{(m,-\infty)}(E^{n+1}_{+})$, $v_j \in \mathcal{H}_{(-\infty)}(R^n_o)$, i.e.

$$(2.6) \qquad (T^{*} V)(u) = \langle u, P^{*}_{\eta} v\rangle + \sum_{l=0}^{m-1} \langle D_n^{m-1-l}u(\cdot,0),\ iP^{*}_{l,\eta}v(\cdot,0) + \sum_{j=1}^{\varkappa} q^{*}_{j,l}v_j\rangle .$$

The completion of the space $C^{\infty}_{(o)}(E^{n+1}_{+})$ in the norm $\|u\|_o$ is isomorphic with the space $\mathcal{H}_{(o,s)}(E^{n+1}_{+}) \oplus \bigoplus_{l=0}^{m-1} \mathcal{H}_{(-l+s)}(R^n_o)$ and the adjoint space is isomorphic with $\mathcal{H}_{(o,-s)}(E^{n+1}_{+}) \oplus \bigoplus_{l=0}^{m-1} \mathcal{H}_{(l-s)}(R^n_o)$, the duality between these spaces is established by an extension of the bilinear form

$$\int_{E_+^{n+1}} u(x)\overline{f(x)}\,dx + \sum_{l=0}^{m-1} \int_{R_0^n} D_n^l u(x',0)\overline{g_l(x')}\,dx', \quad u,f \in C_{(o)}^{\infty}(E_+^{n+1}),$$

$g_l \in C_o^{\infty}(R_o^n)$.

The space $\mathcal{D}_o$ is defined then as the set of all $V \in H^*$, for which there exist $f \in \mathcal{H}_{(o,-s)}(E_+^{n+1})$, $g_l \in \mathcal{H}_{(l-s)}(R_o^n)$ such that

$$(2.7) \quad (T^* V)(u) = \langle u,f \rangle + \sum_{l=0}^{m-1} \langle D_n^l u(\cdot,0), g_l \rangle , \quad u \in C_o^{\infty}(E_+^{n+1}) .$$

From (2.4) we have

$$\langle P_\eta u,v \rangle = \langle u,f \rangle \quad \text{for} \quad u \in C_o^{\infty}(E_+^{n+1}) ,$$

hence $P_\eta^* v = f$ in E_+^{n+1} and Theorem 4.3.1 from [4] implies that $v \in \mathcal{H}_{(m,-s-m)}(E_+^{n+1})$. Using the equalities (2.6) and (2.7) we obtain

$$\langle u,P_\eta^* v \rangle + \sum_{l=0}^{m-1} \langle D_n^{m-1-l} u(\cdot,0),\ iP_{l,\eta}^* v(\cdot,0) + \sum_{j=1}^{\varkappa} q_{j,l}^* v_j \rangle =$$

$$= \langle u,f \rangle + \sum_{l=0}^{m-1} \langle D_n^{m-1-l} u(\cdot,0) ,\ g_{m-1-l} \rangle \text{ for all } u \in C_{(o)}^{\infty}(E_+^{n+1}),$$

thus:

$$(2.8) \quad P_\eta^* v = f , \quad iP_{l,\eta}^* v(\cdot,0) + \sum_{j=1}^{\varkappa} q_{j,l}^* v_j = g_{m-1-l} ,$$

$l = 0,1,\dots,m-1$,

where $f \in \mathcal{H}_{(o,-s)}(E_+^{n+1})$, $v \in \mathcal{H}_{(m,-s-m)}(E_+^{n+1})$,

$$g_{m-1-l} \in \mathcal{H}_{(m-1-l-s)}(R^n_o), \quad v_j \in \mathcal{H}_{(m_j-m+1-s)}(R^n_o).$$

From (2.7) and the definition of the norm $\|u\|_o$ we have the following equality for the right hand side of (2.5):

$$(2.9) \quad \sup_{u\in C^\infty_{(o)}(E^{n+1}_+)} \frac{|(T^*V)(u)|}{\|u\|_o} = \left(\|f\|^2_{(o,-s),\eta} + \sum_{l=0}^{m-1} \|g_l\|^2_{(l-s),\eta} \right)^{\frac{1}{2}}.$$

Furthermore, the operator T^* restricted to the domain $\mathcal{D}_o$ is represented by

$$(2.10) \quad T_o : \mathcal{D}_o \ni V = (v, \{v_j,\ j=1,\dots,\varkappa\}) \longrightarrow$$
$$\longrightarrow (P^*_\eta v, \{iP^*_{l,\eta} v(\cdot,0) + \sum_{j=1}^{\varkappa} q^*_{j,l} v_j,\quad l=0,1,\dots,m-1\}) \in$$
$$\in \mathcal{H}_{(o,-s)}(E^{n+1}_+) \oplus \bigoplus_{l=0}^{m-1} \mathcal{H}_{(m-1-l-s)}(R^n_o)$$

and from (2.9) we see that the inequality (2.5) takes the form:

$$(2.11) \quad \left(|||v|||^2_{(m-1,-s),\eta} + \sum_{j=1}^{\varkappa} \|v_j\|^2_{(m_j-s),\eta} \right)^{\frac{1}{2}} \leq$$
$$\leq K \left(\|f\|^2_{(o,-s),\eta} + \sum_{l=0}^{m-1} \|g_l\|^2_{(l-s),\eta} \right)^{\frac{1}{2}}.$$

The symbols of the main parts of the operators P^{*}_{η} , $P^{*}_{1,\eta}$ and $q^{*}_{j,1}$ are equal to $\overline{P^{o}}(x, \zeta', \zeta_n)$, $\overline{P^{o}_{1}}(x', \zeta', \zeta_n)$ and $\overline{q_{j,1}(x', \zeta')}$ respectively, where the dashes over tha polynomials P^{o} and P^{o}_{1} in ζ_n denote that the complex conjugate coefficients are taken.

Now we construct two systems of functions :
$\left\{e_{k,1}(x', \zeta'),\ k = 1,\dots,\varkappa,\ \ 1 = 0,1,\dots,m-1\right\}$ and
$\left\{f_{k,1}(x', \zeta'),\ k = 1,\dots,m-\varkappa,\ \ 1 = 0,1,\dots,m-1\right\}$, such that $e_{k,1}, f_{k,1} \in C^{\infty}(R^{n}_{0} \times \overline{Z^{n+1}_{-}} \setminus \{0\})$, $e_{k,1}$ and $f_{k,1}$ are homogeneous in ζ' of degree m-1-l, and

$$(2.12) \qquad \sum_{l=0}^{m-1} e_{k,1}(x', \zeta')\overline{q_{j,1}(x', \zeta')} = |\zeta'|^{m_j}\delta_{k,j},$$

$k,j = 1,\dots,\varkappa$,

$$(2.13) \qquad \sum_{l=0}^{m-1} f_{k,1}(x', \zeta')\overline{q_{j,1}(x', \zeta')} = 0,$$

$k = 1,\dots,m-\varkappa$, $j = 1,\dots,\varkappa$.

Let $\overline{S^{n}_{+}}$ denotes the closed hemisphere $\left\{\zeta' \in \overline{Z^{n+1}_{-}} : |\zeta'| = 1\right\}$ and let V_m be the vector space of all polynomials in ζ_n of degree $< m$. Assume that a hermitian structure is added to V_m. Let $Q_{x',\zeta'}$ denotes $\varkappa$-dimensional subspace of V_m spanned by the polynomials $\overline{Q_j}(x', \zeta, \zeta_n)$, $j = 1,\dots,\varkappa$, and let $R_{x',\zeta'}$ denotes the orthogonal complementation of

$Q_{x',\zeta'}$ in V_m. The subspaces $Q_{x',\zeta'}$ and $R_{x',\zeta'}$ are fibres of smooth vector bundles on $R^n_o \times \overline{S^n_+}$. Since the space $R^n_o \times \overline{S^n_+}$ is contractible, these vector bundles are trivial. Choose a basis $\{R_j(x',\zeta',\zeta_n),\ j = 1,\dots,m-\varkappa\}$ in each fibre $R_{x',\zeta'}$, smoothly depending on x', ζ'. Because the coefficients of the polynomials $\overline{Q_j}$ are constant beyond some compact set, R_j may be chosen also with coefficients constant beyond some compact set. Let V^*_m denotes the adjoint space to V_m, and if U is a subspace of V_m, then $U^{\perp} = \{f \in V^*_m : f(U) = \{0\}\}$.

Take the basis $\{e^k_{x',\zeta'},\ k = 1,\dots,\varkappa,\ f^k_{x',\zeta'},\ k = 1,\dots,m-\varkappa\}$ in V^*_m dual to the basis $\{\overline{Q_j},\ j = 1,\dots,\varkappa,\ R_j,\ j=1,\dots,m-\varkappa\}$ in V_m. Thus we have :

$$e^k_{x',\zeta'}(\overline{Q_j}(x',\zeta',\zeta_n)) = \delta_{k,j},\quad k,j = 1,\dots,\varkappa,$$

$$e^k_{x',\zeta'}(R_j(x',\zeta',\zeta_n)) = 0,\quad j = 1,\dots,m-\varkappa,\quad k = 1,\dots,\varkappa,$$

$$f^k_{x',\zeta'}(\overline{Q_j}(x',\zeta',\zeta_n)) = 0,\quad j = 1,\dots,\varkappa,\quad k = 1,\dots,m-\varkappa,$$

$$f^k_{x',\zeta'}(R_j(x',\zeta',\zeta_n)) = \delta_{k,j},\quad k,j = 1,\dots,m-\varkappa,$$

$\{e^k_{x',\zeta'},\ k = 1,\dots,\varkappa\}$ and $\{f^k_{x',\zeta'},\ k = 1,\dots,m-\varkappa\}$ are bases in $R^{\perp}_{x',\zeta'}$ and $Q^{\perp}_{x',\zeta'}$ respectively, and $e^{k}_{x',\zeta'}$, $f^{k}_{x',\zeta'}$ depend smoothly on $(x',\zeta') \in R^n_o \times \overline{S^n_+}$. Putting

$$e_{k,l}(x',\zeta') = e^{k}_{x',\zeta'}(\zeta_n^{\,m-1-l}), \quad k=1,\ldots,\varkappa, \quad l=0,1,\ldots,m-1,$$

$$f_{k,l}(x',\zeta') = f^{k}_{x',\zeta'}(\zeta_n^{m-1-l}), \quad k=1,\ldots,m-\varkappa, \quad l=0,1,\ldots,m-1$$

and extending these functions on the set $R^n_o \times (\overline{Z^{n+1}_-} \setminus \{0\})$ to homogeneous functions in ζ', each of them of degree m-1-l respectively, we obtain the required systems satisfying the conditions (2.12) and (2.13).

In the sequel it will be important to consider the pseudodifferential operators with the symbols $e_{k,l}(x',\zeta')$ and $f_{k,l}(x',\zeta')$, and the operators $S_k(x',D+i\eta)$ of order m-1, given by the symbols

$$S_k(x',\zeta',\zeta_n) = \sum_{l=0}^{m-1} f_{k,l}(x',\zeta')\overline{P^o_l}(x',\zeta',\zeta_n), \quad k=1,\ldots,m-\varkappa.$$

We shall investigate more closely the properties of these operators.

Let us recall that

$$\overline{P^o}(x',\zeta',\zeta_n) = P^o(x',\overline{\zeta'},\zeta_n) = Q^+(x',\overline{\zeta'},\zeta_n)Q^-(x',\overline{\zeta'},\zeta_n)$$

and the roots of the polynomials $Q^+(x',\overline{\zeta'},\zeta_n)$ and $Q^-(x',\overline{\zeta'},\zeta_n)$ there are in the upper and lower halfplanes respectively for $\eta_0 < 0$. We have also

$$\overline{P^0}(x', \zeta', \zeta_n) = \overline{Q^+}(x', \zeta', \zeta_n)\overline{Q^-}(x', \zeta', \zeta_n)$$

and the roots of the polynomial $\overline{Q^-}(x', \zeta', \zeta_n)$ lay in the upper halfplane, the roots of $\overline{Q^+}(x', \zeta', \zeta_n)$ in the lower halfplane, hence

$$Q^+(x', \overline{\zeta'}, \zeta_n) = \overline{Q^-}(x', \zeta', \zeta_n) , \quad Q^-(x', \overline{\zeta'}, \zeta_n) = \overline{Q^+}(x', \zeta', \zeta_n).$$

From this we infer that the polynomial $Q^+(x', \overline{\zeta'}, \zeta_n)$ is of degree $m-\varkappa$ and the polynomials $\overline{Q_j}(x', \zeta', \zeta_n)$, $j=1,\ldots,\varkappa$ are independent $\mathrm{mod}\, Q^-(x', \overline{\zeta'}, \zeta_n)$. The last statement may be reformulated as follows: each nonzero polynomial from the space $Q_{x', \zeta'}$ is not divisible by the polynomial $Q^-(x', \overline{\zeta'}, \zeta_n)$.

Now we are going to prove that the polynomials $S_k(x', \zeta', \zeta_n)$, $k = 1,\ldots,m-\varkappa$, are linearly independent $\mathrm{mod}\, Q^+(x', \overline{\zeta'}, \zeta_n)$ for $(x', \zeta') \in R^n_0 \times (\overline{Z^{n+1}_+} \setminus \{0\})$.
By the above we should prove that each nonzero polynomial from the space spanned by the polynomials $S_k(x', \zeta', \zeta_n)$, $k = 1,\ldots,m-\varkappa$, is not divisible by $Q^+(x', \overline{\zeta'}, \zeta_n)$. It is a consequence of the following lemma.

Lemma. Let V_m denotes m-dimensional complex vector space of polynomials in z of degree $< m$, V^*_m - the adjoint space,

$$p(z) = z^m + a_1 z^{m-1} + \dots + a_m = q^+(z) q^-(z) ,$$

$$q^+(z) = z^{m-\varkappa} + b_1 z^{m-\varkappa-1} + \dots + b_{m-\varkappa} ,$$

$$q^-(z) = z^{\varkappa} + c_1 z^{\varkappa-1} + \dots + c_{\varkappa} ,$$

$Q \subset V_m$ is $\varkappa$-dimensional linear subspace in which every nonzero element is not divisible by $q^-(z)$.

Let us denote :

$$p_k(z) = z^k + a_1 z^{k-1} + \dots + a_k , \quad k = 0,1,\dots,m-1 ,$$

$$q_k^+(z) = z^k + b_1 z^{k-1} + \dots + b_k, \quad k = 0,1,\dots,m-\varkappa ,$$

$$q_k^-(z) = z^k + c_1 z^{k-1} + \dots + c_k, \quad k = 0,1,\dots,\varkappa ,$$

thus $p_o(z) = q_o^+(z) = q_o^-(z) = 1$, $q_{m-\varkappa}^+(z) = q^+(z)$,

$q_{\varkappa}^-(z) = q^-(z)$.

Define the transformation $i : V_m^* \longrightarrow V_m$ by

$$(if)(z) = \sum_{k=0}^{m-1} f(z^{m-1-k}) p_k(z) , \qquad f \in V_m .$$

Let $Q^{\perp} = \left\{ f \in V_m^* : f(Q) = \{0\} \right\}$.

Then each nonzero element of the $m-\varkappa$-dimensional subspace $i(Q^{\perp})$ is not divisible by $q^+(z)$.

Proof. The transformation i is a homomorphism.

Let $e_k(z) = z^k$, $k = 0,1,\dots,\varkappa-1$, $e_k(z) = z^{k-\varkappa} q^-(z)$,

$k = \varkappa, \dots, m-1$.

The system $\{e_k(z)\ ,\ k=0,1,\dots,m-1\}$ is a basis in V_m .

The dual basis $\{f_k \in V_m^*\ ,\ k = 0,1,\dots,m-1\}$ is defined by the equalities

$$f_j(e_k) = \delta_{j,k}, \qquad j,k = 0,1,\dots,m-1 .$$

Now we shall find the image of the basis $\{f_k,\ k=0,1,\dots,m-1\}$ by i.

$$z^j = \sum_{k=0}^{m-1} a_{j,k} e_k(z)\ , \qquad j = 0,1,\dots,m-1 .$$

Since $f_k(z^j) = a_{j,k}$, by the properties of the polynomials $e_k(z)$ we get :

$a_{j,k} = 0$ for $j < k$, $a_{j,j} = 1$, $a_{j,k} = 0$ for $k < j < \varkappa$.

Thus the matrix $A = (a_{j,k})$, $j,k = 0,1,\dots,m-1$, is of the form

$$A = \left(\begin{array}{ccc|cccc} 1 & & 0 & & & 0 & \\ & 1 & & & & & \\ & & \ddots & & & & \\ 0 & & & & & & \\ \hline a_{\varkappa 0} & \dots & a_{\varkappa,\varkappa-1} & 1 & & 0 & \\ a_{\varkappa+1,0} & \dots & a_{\varkappa+1,\varkappa-1} & a_{\varkappa+1,\varkappa} & 1 & & \\ \vdots & & \vdots & \vdots & \vdots & \ddots & \\ a_{m-1,0} & \dots & a_{m-1,\varkappa-1} & a_{m-1,\varkappa} & a_{m-1,\varkappa+1} & \dots & 1 \end{array}\right) .$$

Gathering in the sum $\sum_{k=0}^{m-1} a_{j,k} e_k(z)$ terms at the same powers of z and comparing them to z^j we get:

$$(2.14)\qquad a_{j,k} + \sum_{l=\varkappa}^{\varkappa+k} a_{j,l}c_{l-k} = 0 \quad \text{for } j \geq \varkappa,\ k=0,1,\dots,\varkappa-1,$$

$$(2.15)\qquad \sum_{l=k}^{j} a_{j,l}c_{l-k} = 0 \quad \text{for } j \geq \varkappa,\ k=\varkappa,\dots,j-1,$$

where $c_l = 0$ for $l > \varkappa$ and $c_o = 1$.

Analogously by the equality $p(z) = q^+(z)q^-(z)$ we get the following formulas for $p_k(z)$:

$$(2.16)\qquad p_k(z) = q_k^+(z) + c_1 q_{k-1}^+(z) + \dots + c_k q_o^+(z),$$

$k = 0,1,\dots,m-\varkappa-1$,

$$(2.17)\qquad p_k(z) = q^-_{k-(m-\varkappa)}(z)q^+(z) + c_{k-(m-\varkappa)+1}q^+_{m-\varkappa-1}(z)+\dots+$$

$$+ c_\varkappa q^+_{k-\varkappa}(z), \quad k = m-\varkappa,\dots,m-1,$$

where for $k < 0$ $q_k^+(z)$ denotes the zero polynomial.

By induction we shall prove that

$$(2.18)\qquad \sum_{j=k}^{m-1} a_{j,k}p_{m-1-j}(z) = q^+_{m-1-k}(z), \quad k=\varkappa,\dots,m-1.$$

For $k=m-1$ it is obvious. Assume the validity of the formula (2.18) for $k+1,\dots,m-1$, where $k \geq \varkappa$. We shall prove it for k. By (2.15)

$$\sum_{l=k}^{j} a_{j,l}c_{l-k} = 0, \quad j = k+1,\dots,m-1,$$

multiplying these equalities by $p_{m-1-j}(z)$, adding all of them and adding to both sides of the obtained equality $p_{m-1-k}(z)$ we get

$$p_{m-1-k}(z) = \sum_{j=k}^{m-1} a_{j,k}p_{m-1-j}(z) + \sum_{l=k+1}^{m-1} c_{l-k}\sum_{j=l}^{m-1} a_{j,l}p_{m-1-j}(z) .$$

Now, by the induction assumption and by (2.16) it is

$$q^+_{m-1-k}(z) + \sum_{l=k+1}^{m-1} c_{l-k}q^+_{m-1-l}(z) = \sum_{j=k}^{m-1} a_{j,k}p_{m-1-j}(z) + \sum_{l=k+1}^{m-1} c_{l-k}q^+_{m-1-l}(z) ,$$

hence the induction is completed.

The next step is to prove that

$$(2.19)\quad \sum_{j=k}^{m-1} a_{j,k}p_{m-1-j}(z) = q^-_{\varkappa-1-k}(z)q^+(z) , \quad k=0,\ldots,\varkappa-1.$$

From (2.14) by computation as above we have

$$p_{m-1-k}(z) = \sum_{j=k}^{m-1} a_{j,k}p_{m-1-j}(z) + \sum_{l=\varkappa}^{\varkappa+k} c_{l-k} \sum_{j=l}^{m-1} a_{j,l}p_{m-1-j}(z) .$$

Using (2.17) and (2.18) we have

$$q^-_{\varkappa-1-k}(z)q^+(z) + \sum_{l=\varkappa}^{\varkappa+k} c_{l-k}q^+_{m-1-l}(z) = \sum_{j=k}^{m-1} a_{j,k}p_{m-1-j}(z) + \sum_{l=\varkappa}^{\varkappa+k} c_{l-k}q^+_{m-1-l}(z) ,$$

and hence (2.19).

This gives formulas for if_k, because

$$(if_k)(z) = \sum_{j=0}^{m-1} f_k(z^j)p_{m-1-j}(z) = \sum_{j=k}^{m-1} a_{j,k}p_{m-1-j}(z) \ , \quad \text{or}$$

$$(if_k)(z) = q^-_{\varkappa-1-k}(z)q^+(z) \ , \qquad k = 0,\ldots,\varkappa-1 \ ,$$

$$(if_k)(z) = q^+_{m-1-k}(z) \ , \qquad k = \varkappa,\ldots,m-1 \ .$$

The system of polynomials $\{(if_k)(z) \ , \ k = 0,1,\ldots,m-1\}$ is a basis in the space V_m and this implies that i is an isomorphism.

In the space Q choose a basis $\{q_j(z), \ j = 0,1,\ldots,\varkappa-1\}$.

Let

$$q_j(z) = \sum_{k=0}^{m-1} d_{j,k}e_k(z) \ , \qquad j = 0,1,\ldots,\varkappa-1 \ .$$

By the properties of Q and of the system $\{e_k(z), \ k=0,1,\ldots,m-1\}$ it is

$$\det(d_{j,k})_{j,k=0}^{\varkappa-1} \neq 0 \ . \tag{2.20}$$

Now it is seen how to end the proof of the lemma.

Let $f \in Q^\perp$, $f \neq 0$, so $f = \sum_{k=0}^{m-1} z_k f_k$. Because $f(q_j) = 0$ we have the following equalities:

$$\sum_{k=0}^{m-1} d_{j,k}z_k = 0 \ , \qquad j = 0,1,\ldots,\varkappa-1 \ .$$

From this we infer that $(z_{\varkappa},\dots,z_{m-1}) \neq 0$. Indeed, on the contrary, if $(z_{\varkappa},\dots,z_{m-1}) = 0$, by (2.20) $z_k = 0$ for $k = 0,1,\dots,m-1$ and this contradicts $f \neq 0$. Because

$$(if)(z) = \sum_{k=0}^{m-1} z_k(if_k)(z) = \left(\sum_{k=0}^{\varkappa-1} z_k q^-_{\varkappa-1-k}(z)\right) q^+(z) + \sum_{k=\varkappa}^{m-1} z_k q^+_{m-1-k}(z), \text{ and}$$

$$\sum_{k=\varkappa}^{m-1} z_k q^+_{m-1-k}(z) \neq 0$$

and it is not divisible by $q^+(z)$, the polynomial $(if)(z)$ is not divisible by $q^+(z)$ also.

Thus the proof of the lemma is completed.

From the considerations we have done it follows that the operators P^*_η and $S_{k,\eta}$, $k = 1,\dots,m-\varkappa$, satisfy the assumptions of Theorem 2.

Applying the pseudodifferential operators $e_{k,l}(x',D'+i\eta')$ and $f_{k,l}(x',D'+i\eta')$ to the equalities (2.8) we obtain:

$$(2.21)\qquad P^*_\eta v = f, \quad S_{k,\eta}v(\cdot,0) + T_k v(\cdot,0) + \sum_{j=1}^{\varkappa} s_{k,j}v_j = -i\sum_{l=0}^{m-1} f_{k,l}g_{m-1-l}$$

where $k = 1,\dots,m-\varkappa$, T_k are the operators of order $m-2$, $s_{k,j}$ are of order m_j-1;

(2.22) $\Lambda_{(m_k),\eta} v_k + \sum_{j=1}^{\varkappa} t_{k,j} v_j + R_k v(\cdot,0) = \sum_{l=0}^{m-1} e_{k,l} g_{m-1-l}$,

where $k = 1,\dots,\varkappa$, R_k are the operators of order $m-1$, $t_{k,j}$ are of order m_j-1.

Here we have used the propositions (i) and (ii), the equalities (2.12), (2.13), and the definition of the operators $S_{k,\eta}$.

Now we shall prove that $v \in \mathcal{B}_{(m-1,-s)}(E_+^{n+1})$ and $v_j \in \mathcal{H}_{(m_j-s)}(R_o^n)$, $j = 1,\dots,\varkappa$.

We know that $v \in \mathcal{B}_{(m-1,-s-m)}(E_+^{n+1})$ and $v_j \in \mathcal{H}_{(m_j-s-m)}(R_o^n)$.

Thus it suffices to prove the following statement:

if $v \in \mathcal{B}_{(m-1,\tau)}(E_+^{n+1})$, $v_j \in \mathcal{H}_{(m_j+\tau)}(R_o^n)$ and $\tau + 1 \leq -s$, then $v \in \mathcal{B}_{(m-1,\tau+1)}(E_+^{n+1})$ and $v_j \in \mathcal{H}_{(m_j+\tau+1)}(R_o^n)$.

Let v and v_j satisfy the assumptions of this statement.

Because $-i\sum_{l=0}^{m-1} f_{k,l} g_{m-1-l} \in \mathcal{H}_{(-s)}(R_o^n)$, $s_{k,j} v_j \in \mathcal{H}_{(\tau+1)}(R_o^n)$ and $T_k v(\cdot,0) \in \mathcal{H}_{(\tau+1)}(R_o^n)$, $f \in \mathcal{H}_{(o,-s)}(E_+^{n+1})$, hence from (2.21) $P_\eta^* v \in \mathcal{H}_{(o,\tau+1)}(E_+^{n+1})$, $S_{k,\eta} v(\cdot,0) \in \mathcal{H}_{(\tau+1)}(R_o^n)$,

and applying Theorem 2 we obtain $v \in \mathcal{B}_{(m-1,\tau+1)}(E_+^{n+1})$.

Then from (2.22) by the similar consideration we have

$$\Lambda_{(m_k),\eta} v_k \in \mathcal{H}_{(\tau+1)}(R^n_o)\ , \text{ hence } \ v_k \in \mathcal{H}_{(m_k+\tau+1)}(R^n_o)$$

and the statement is proved.

The last task is to prove the inequality (2.11). By Theorem 2 and (2.21) we have:

$$|||v|||^2_{(m-1,-s),\eta} \le K\Big(\|f\|^2_{(o,-s),\eta} + \sum_{l=0}^{m-1} \|g_l\|^2_{(l-s),\eta} + |||v|||^2_{(m-2,-s),\eta} + \sum_{j=1}^{\varkappa} \|v_j\|^2_{(m_j-1-s),\eta}\Big),$$

hence this inequality holds also for $|\eta|$ sufficiently large if the term $|||v|||^2_{(m-2,-s),\eta}$ is rejected. From (2.22) and the above inequality:

$$\sum_{j=1}^{\varkappa} \|v_j\|^2_{(m_j-s),\eta} \le K\Big(\|f\|^2_{(o,-s),\eta} + \sum_{l=0}^{m-1} \|g_l\|^2_{(l-s),\eta} + \sum_{j=1}^{\varkappa} \|v_j\|^2_{(m_j-1-s),\eta}\Big),$$

hence again we have the inequality without the last term on the right hand side. Combining these two inequalities we get (2.11). Thus the Theorem 3 is proved.

Similarly as in the first section we shall generalize this theorem to the case of arbitrary nonnegative integers m_j. Let us recall the denotations $m_o = \max_{1 \le j \le \varkappa} m_j$ and introduce the norm

$$\|u\|_o = \Big(\|u\|^2_{(o,k+s),\eta} + \sum_{l=0}^{k-1} \|D_n^l \, P_\eta u(\cdot,0)\|^2_{(-m-l+k+s),\eta} +$$

$$+ \sum_{l=0}^{m-1} \|D_n^l u(\cdot,0)\|^2_{(-l+k+s),\eta} \Big)^{\frac{1}{2}}, \quad u \in C^\infty_{(o)}(E^{n+1}_+).$$

By $\mathcal{D}_o$ we shall denote now the set of all $V \in H^*$, i.e. $v \in \mathring{\mathcal{H}}_{(-k,-s)}(\overline{E^{n+1}_+})$ and $v_j \in \mathcal{H}_{(-m+1-k+m_j-s)}(R^n_o)$, for which the functional $(T^* V)(u)$ is continuous in the norm $\|u\|_o$. The operator T^* is considered in the domain $\mathcal{D}_o$. Since we are not interested in the regularity results for the elements of $\mathcal{D}_o$, therefore we restrict ourselves to the following formulation.

Theorem 4. If the assumptions of Theorem 2 are satisfied for the operators $P(x,D+i\eta)$ and $Q_j(x',D+i\eta)$, $j=1,\dots,\varkappa$, k is a nonnegative integer such that $m-1+k \geq m_o$, and s is arbitrary real number, then there exists a constant K such that for all $v \in \breve{\mathcal{H}}_{(-k,-s)}(\overline{E^{n+1}_+})$, $v_j \in \mathcal{H}_{(-m+1-k+m_j-s)}(R^n_o)$, $j = 1,\dots,\varkappa$, $|\eta|$ sufficiently large, $\eta_o < 0$, there holds inequality

$$(2.29) \quad \Big(\|v\|^2_{(-k,-s),\eta} + \sum_{j=1}^{\varkappa} \|v_j\|^2_{(-m+1-k+m_j-s),\eta} \Big)^{\frac{1}{2}} \leq$$

$$\leq K \sup_{u \in C^\infty_{(o)}(E^{n+1}_+)} \frac{|(T^* V)(u)|}{\|u\|_o} .$$

Proof. Of course it is sufficient to prove this inequality for $V \in \mathcal{D}_0$. Its right hand side can be estimated from below by $\sup_{u \in C_0^\infty(E_+^{n+1})} \frac{|(T^* V)(u)|}{\|u\|_0}$, but this expression is equal to $\sup_{u \in C_0^\infty(E_+^{n+1})} \frac{|\langle P_\eta u, v\rangle|}{\|u\|_{(0,k+s),\eta}}$. Because the space $C_0^\infty(E_+^{n+1})$ is dense in $\mathcal{H}_{(0,k+s)}(E_+^{n+1})$ it is equal also to $\|P_\eta^* v\|_{(0,-k-s),\eta}$. Thus, denoting by $v|_{E_+^{n+1}}$ the restriction of the distribution v to the domain E_+^{n+1}, we have $P_\eta^* v|_{E_+^{n+1}} \in \mathcal{H}_{(0,-k-s)}(E_+^{n+1})$, hence, by Theorem 4.3.1 of [4], $v|_{E_+^{n+1}} \in \mathcal{H}_{(m,-k-s-m)}(E_+^{n+1})$. Introduce the distribution v_0: $v_0 = v|_{E_+^{n+1}}$ in E_+^{n+1} and $v_0 = 0$ in $\overline{E_-^{n+1}}$. Then $v_0 \in \mathring{\mathcal{H}}_{(0,-k-s)}(\overline{E_+^{n+1}})$, hence also $v_0 \in \mathring{\mathcal{H}}_{(-k,s)}(\overline{E_+^{n+1}})$, and $\operatorname{supp}(v-v_0) \subset R_0^n$. Such distribution is of the form

$$v - v_0 = \sum_{l=0}^{k-1} f_l(x') \otimes D_n^l \delta(x_n),$$

where $f_l \in \mathcal{H}_{(-k+l+\frac{1}{2}-s)}(R_0^n)$.

Taking into account the properties of the distribution v we can integrate by parts in the formula (2.4). To do it let us notice that

$$Q_{j,\eta} = R_j P_\eta + Q'_{j,\eta},$$

where $R_j = \sum_{l=0}^{k-1} r_{j,l} D_n^l$, $Q'_{j,\eta} = \sum_{l=0}^{m-1} q_{j,l} D_n^{m-1-l}$ and $r_{j,l}$, $q_{j,l}$ are the operators of order m_j-m-l, $m_j+l-m+1$ respectively, acting on the distributions defined on R_o^n. They are compositions of pseudodifferential and differential operators. Then

$$(T^* V)(u) = \langle u, P_\eta^* v_o\rangle + \sum_{l=0}^{k-1} \langle D_n^l P_\eta u(\cdot,0), f_l + \sum_{j=1}^{\varkappa} r_{j,l}^* v_j\rangle +$$

$$+ \sum_{l=0}^{m-1} \langle D_n^{m-1-l} u(\cdot,0),\ i P_{l,\eta}^* v_o(\cdot,0) + \sum_{j=1}^{\varkappa} q_{j,l}^* v_j\rangle,$$

and, by the definition of the norm $\|u\|_o$,

$$(2.24) \qquad \sup_{u \in C_{(o)}^\infty(E_+^{n+1})} \frac{|(T^* V)(u)|}{\|u\|_o} =$$

$$= \Big(\|P_\eta^* v_o\|^2_{(o,-k-s),\eta} + \sum_{l=0}^{k-1} \|f_l + \sum_{j=1}^{\varkappa} r_{j,l}^* v_j\|^2_{(m+l-k-s),\eta} +$$

$$+ \sum_{l=0}^{m-1} \|i P_{l,\eta}^* v_o(\cdot,0) + \sum_{j=1}^{\varkappa} q_{j,l}^* v_j\|^2_{(m-1-l-k-s),\eta} \Big)^{\frac{1}{2}}.$$

The operators P_η and $\Lambda_{(m-1-m_j),\eta} Q'_{j,\eta}$ satisfy the assumptions of Theorem 3, so it follows from (2.11) and (2.8) that:

$$(2.25)\qquad \| P^{*}_{\eta} v_o \|^2_{(o,-k-s),\eta} + \sum_{l=0}^{m-1} \| iP^{*}_{l,\eta} v_o(\cdot,0) + \sum_{j=1}^{\varkappa} (q^{*}_{j,l} \Lambda_{(m-1-m_j),\eta}) \Lambda_{(m_j-m+1),\eta} v_j \|^2_{(m-1-l-k-s),\eta} \geq$$

$$\geq K^{-2} (|||v_o|||^2_{(m-1,-k-s),\eta} + \sum_{j=1}^{\varkappa} \|v_j\|^2_{(m_j-k-s),\eta}).$$

In particular it follows that $v_j \in \mathcal{H}_{(m_j-k-s)}(R^n_o)$ and because $r^{*}_{j,l} v_j \in \mathcal{H}_{(m+l-k-s)}(R^n_o)$ and $f_l + \sum_{j=1}^{\varkappa} r^{*}_{j,l} v_j \in \mathcal{H}_{(m+l-k-s)}(R^n_o)$, we have $f_l \in \mathcal{H}_{(m+l-k-s)}(R^n_o)$ also.

Combining (2.24), (2.25) and estimating

$$\| f_l + \sum_{j=1}^{\varkappa} r^{*}_{j,l} v_j \|^2_{(m+l-k-s),\eta} \geq$$

$$\geq \frac{1}{2} \| f_l \|^2_{(m+l-k-s),\eta} - K_1 \sum_{j=1}^{\varkappa} \| v_j \|^2_{(m_j-k-s),\eta}$$

we obtain the inequality

$$(2.26)\qquad \sup_{u \in C^{\infty}_{(o)}(E^{n+1}_+)} \frac{|(T^{*} V)(u)|}{\|u\|_o} \geq K^{-1} \Big(|||v_o|||^2_{(m-1,-k-s),\eta} + \varepsilon \sum_{l=0}^{k-1} \|f_l\|^2_{(m+l-k-s),\eta} + (1 - 2\varepsilon K_1) \sum_{j=1}^{\varkappa} \|v_j\|^2_{(m_j-k-s),\eta} \Big)^{\frac{1}{2}},$$

in which $\varepsilon > 0$ is chosen such that $1 - 2\varepsilon K_1 > 0$.

Now we can estimate v in the norm of $\mathring{\mathcal{H}}_{(-k,-s)}(\overline{E^{n+1}_+})$:

$$\|v\|^2_{(-k,-s),\eta} = \Big\|v_0 + \sum_{l=0}^{k-1} f_l \otimes D_n^l \delta\Big\|^2_{(-k,-s),\eta} \leq$$

$$\leq K_0\Big(\|v_0\|^2_{(-k,-s),\eta} + \sum_{l=0}^{k-1} \|f_l\|^2_{(l+\frac{1}{2}-k-s),\eta}\Big) \leq$$

$$\leq K_0\Big(\|v_0\|^2_{(m-1,-k-s),\eta} + \sum_{l=0}^{k-1}\|f_l\|^2_{(m+l-k-s),\eta}\Big) .$$

From this estimation and (2.26) we get (2.23) with a suitable constant K.

3. The existence theorems.

In this section the results obtained in Theorem 2 and Theorem 4 are applied to differential boundary operators $Q_j(x',D)$, $j = 1,\dots,\varkappa$. We assume that they are of order m_j respectively and their coefficients are smooth functions constant outside a compact subset of R^n_0 .

Now we are inerested in solving the equations

$$P(x,D)u = f , \quad Q_j(x',D)u(x',0) = g_j, \quad j=1,\dots,\varkappa ,$$

which are not the equations of the preceding form, i.e.

$$P(x,D+i\eta)u = f, \quad Q_j(x',D+i\eta)u(x',0) = g_j, \quad j=1,\dots,\varkappa,$$

hence we modify the definitions and theorems obtained above.

Because $e^{\langle x,\eta\rangle}P(x,D)u(x) = P(x,D+i\eta)(e^{\langle x,\eta\rangle}u(x))$ and, similarly, $e^{\langle x',\eta'\rangle}Q_j(x',D)u(x',0) = Q_j(x',D+i\eta)(e^{\langle x',\eta'\rangle}u(x',0))$ it is easily seen how to do these modifications.

Thus, by $\mathcal{H}_{(s),\eta}(R^n)$ we shall mean the space of all $u\in\mathcal{D}'(R^n)$ such that $e^{\langle x',\eta'\rangle}u \in \mathcal{H}_{(s)}(R^n)$, with the norm

$$\|u\|^2_{(s),\eta} = (2\pi)^{-n}\int |\zeta'|^{2s}|\hat{u}(\zeta')|^2 d\xi' .$$

Similarly, by $\mathcal{H}_{(\tau,s),\eta}(R^{n+1})$ we shall denote the space of all $u \in \mathcal{D}'(R^{n+1})$ such that $e^{\langle x,\eta\rangle}u \in \mathcal{H}_{(\tau,s)}(R^{n+1})$, with the norm

$$\|u\|^2_{(\tau,s)} = (2\pi)^{-n-1}\int |\zeta|^{2\tau}|\zeta'|^{2s}|\hat{u}(\zeta)|^2 d\xi .$$

$\mathring{\mathcal{H}}_{(\tau,s),\eta}(\overline{E^{n+1}_+})$ is defined as a subspace of $\mathcal{H}_{(\tau,s),\eta}(R^{n+1})$ consisting of all u with $\operatorname{supp} u \subset \overline{E^{n+1}_+}$. In the sequel the symbol $\|u\|_{(\tau,s),\eta}$ will denote the norm in the above space only when τ will be a negative integer.

$\mathcal{H}_{(\tau,s),\eta}(E^{n+1}_+)$ denotes the space of all $u \in \mathcal{D}'(E^{n+1}_+)$ which have extensions $U\in \mathcal{H}_{(\tau,s),\eta}(R^{n+1})$, equipped with the quotient norm

$$\|u\|_{(\tau,s),\eta} = \inf_U \|U\|_{(\tau,s),\eta} .$$

where the infimum is taken with respect to all extensions. This norm shall be applied only for τ being a nonnegative integer.

Finally by $\mathcal{B}_{(k,s),\eta}(E_+^{n+1})$, k a nonnegative integer, we mean the space of all $u \in \mathcal{H}_{(k+1,s-1),\eta}(E_+^{n+1})$ such that $D_n^l u(\cdot,0) \in \mathcal{H}_{(k-l+s),\eta}(R_o^n)$, $l = 0,1,\ldots,k$, with the norm

$$|||u|||^2_{(k,s),\eta} = |\eta| \, \|u\|^2_{(k,s),\eta} + \sum_{l=0}^{k} \|D_n^l u(\cdot,0)\|^2_{(k-l+s),\eta} .$$

This space is not complete.

In these definitions we assume that $\eta \neq 0$.

From Theorem 2.5.1 of [4] it follows that the adjoint space to $\mathcal{H}_{(s),\eta}(R^n)$ with respect to the extension of the bilinear form

$$\int_{R^n} u(x)\overline{v(x)}\, dx \quad , \quad u,v \in C_o^\infty(R^n) ,$$

is equal to $\mathcal{H}_{(-s),-\eta}(R^n)$.

The adjoint space to $\mathcal{H}_{(\tau,s),\eta}(E_+^{n+1})$ is equal to $\mathring{\mathcal{H}}_{(-\tau,-s),-\eta}(\overline{E_+^{n+1}})$ and the duality between these spaces is the extension of the bilinear form

$$\int_{E_+^{n+1}} u(x)\overline{v(x)}dx \, , \quad u \in C_{(o)}^\infty(E_+^{n+1}) \, , \quad v \in C_o^\infty(E_+^{n+1}) .$$

The basic result concerning the differential operators immediately follows from Theorems 2 and 4 .

Theorem 5. Let us suppose:

(A) for $x' \in R^n_0$, $\xi' \neq 0$, the real roots of the polynomial $P^0(x', \xi', \zeta_n)$ in ζ_n are at most of double multiplicity;

(B) the polynomials $Q^0_j(x', \zeta', \zeta_n)$ are linearly independent mod $Q^+(x', \zeta', \zeta_n)$ for $x' \in R^n_0$, $\zeta' \neq 0$, $\eta_0 \leq 0$.

Let k be a nonnegative integer and $m-1+k \geqslant m_j$ for $j = 1, \ldots, \varkappa$, and let s be any real number.
Then there exists a constant K such that for $|\eta|$ sufficiently large, $\eta_0 < 0$, we have:

$$|||u|||_{(m-1+k,s),\eta} \leq K\left(\|Pu\|^2_{(k,s),\eta} + \sum_{j=1}^{\varkappa} \|Q_j u(\cdot,0)\|^2_{(m-1+k-m_j+s),\eta}\right)^{\frac{1}{2}}$$

for all $u \in \mathcal{B}_{(m-1+k,s),\eta}(E^{n+1}_+)$;

$$\left(\|v\|^2_{(-k,-s),-\eta} + \sum_{j=1}^{\varkappa} \|v_j\|^2_{(-m+1-k+m_j-s),-\eta}\right)^{\frac{1}{2}} \leq$$

$$\leq K \sup_{u \in C^\infty_{(0)}(E^{n+1}_+)} \frac{\left|\langle Pu, v\rangle + \sum_{j=1}^{\varkappa} \langle Q_j u(\cdot,0), v_j\rangle\right|}{\|u\|_0}$$

for all $v \in \mathring{\mathcal{H}}_{(-k,-s),-\eta}(\overline{E^{n+1}_+})$, $v_j \in \mathcal{H}_{(-m+1-k+m_j-s),-\eta}(R^n_0)$,

where

$$\|u\|^2_0 = \|u\|^2_{(0,k+s),\eta} + \sum_{l=0}^{k-1} \|D^l_n Pu(\cdot,0)\|^2_{(-m-l+k+s),\eta} +$$

$$+ \sum_{l=0}^{m-1} \|D^l_n u(\cdot,0)\|^2_{(-l+k+s),\eta}, \qquad u \in C^\infty_{(0)}(E^{n+1}_+).$$

This theorem implies the existence theorem for the boundary problem in the spaces of functions rapidly decreasing for $x_o \longrightarrow -\infty$. It can be treated as a mixed problem with Cauchy data posed on the hyperplane $x_o = -\infty$.

Theorem 6. Let operators P and Q_j be the same as in Theorem 5. Then for each $f \in \mathcal{H}_{(k,s),\eta}(E_+^{n+1})$, $g_j \in \mathcal{H}_{(m-1+k-m_j+s)}(R_o^n)$, $j = 1,\dots,\varkappa$, and for sufficiently large $|\eta|$ there is exactly one solution $u \in \mathcal{B}_{(m-1+k,s),\eta}(E_+^{n+1})$ of the problem $Pu = f$, $Q_j u(\cdot,0) = g_j$. If f and g_j are in the spaces correspondin g to η^1 and η^2 respectively, then the solutions in the spaces $\mathcal{B}_{(m-1+k,s),\eta}(E_+^{n+1})$ for $\eta = \eta^1$ and $\eta = \eta^2$ are equal.

Proof. Recall the denotations introduced at the begining of this section. From Theorem 5 we conclude that $T\mathcal{D}_T$ is a closed linear subspace of H and each functional from H^* vanishing on $T\mathcal{D}_T$ vanishes identically. Hence $T\mathcal{D}_T = H$ and the first part of Theorem 6 is proved.

From Theorem 5 it follows also that $TC_{(o)}^{\infty}(E_+^{n+1})$ is dense in H. Let us notice generally that if X_i, $i = 1,2$, are Banach spaces with norms $\|\cdot\|_i$ and X is a linear subspace dense in each of the spaces X_i, then X is dense in the space $X_1 \cap X_2$ with the norm $\|\cdot\|_{1,2} = \max(\|\cdot\|_1, \|\cdot\|_2)$. Indeed, take a linear functional $x^* \in (X_1 \cap X_2)^*$ vanishing on X. This functional is of the form $x^* = x_1^* + x_2^*$, where

$x_i^* \in X_i^*$ hence $x_1^* x = -x_2^* x$ for $x \in X$ and this equality implies that the functionals x_i^* are continuous in the topologies of the both spaces X_i. Thus the equality $x_1^* x = -x_2^* x$ holds for all $x \in X_1 \cap X_2$ and x^* is the zero functional. The theorem of Hahn-Banach implies the required fact.

Now let $(f, \{g_j, j=1,\dots,\varkappa\})$ belong to the spaces H with two different η : η^1 and η^2. Because $TC_{(0)}^\infty(E_+^{n+1})$ is dense in both these spaces the above remark implies that there exists a sequence $\{u_j\}$, $j = 1,2,\dots$, $u_j \in C_{(0)}^\infty(E_+^{n+1})$, such that $Tu_j \longrightarrow (f, \{g_j, j = i,\dots,\varkappa\})$ in both these topologies. Theorem 5 implies that u_j is convergent to the solution of the problem $Pu = f$, $Q_j u(\cdot,0) = g_j$ in the spaces $\mathcal{B}_{(m-1+k,s),\eta}(E_+^{n+1})$ for $\eta = \eta^1$ and $\eta = \eta^2$, and the uniqueness of solutions in these spaces implies the second part of Theorem 6.

These two theorems imply the following

Corollary. Let P, Q_j, f, g_j be the same as in Theorem 6 and let $f = 0$ in E_+^{n+1} for $x_0 < T$, $g_j = 0$ in R_0^n for $x_0 < T$. If $u \in \mathcal{B}_{(m-1+k,s),\eta}(E_+^{n+1})$ is a solution of the problem $Pu = f$, $Q_j u(\cdot,0) = g_j$, then $u = 0$ in E_+^{n+1} for $x_0 < T$.

Proof. We can assume that $T = 0$. Then $(f, \{g_j, j=1,\dots,\varkappa\}) \in$ $\in \mathcal{H}_{(k,s),\eta}(E_+^{n+1}) \oplus \bigoplus_{j=1}^{\varkappa} \mathcal{H}_{(m-1+k-m_j+s),\eta}(R_0^n)$ for all $|\eta|$

sufficiently large with norms uniformly bounded, hence also $|||u|||_{(m-1+k,s),\eta}$ is uniformly bounded for $|\eta|$ large and hence we get Corollary.

To formulate the next theorem we need some other function spaces. Let us denote $H_+ = \{x \in E_+^{n+1} : 0 < x_0 < T\}$, $H_0 = \{x' \in R_0^n : 0 < x_0 < T\}$. If k is a nonnegative integer, then by $\mathring{\mathcal{H}}_{(k)}(H_+)$ we denote the space of all $u \in \mathcal{H}_{(k)}(H_+)$ which can be extended to a function $U \in \mathcal{H}_{(k)}(R^{n+1})$ such that $\operatorname{supp} U \subset \overline{R_+^{n+1}}$. The norm in this space is given by

$$\|u\|_{(k)}^2 = \int_{H_+} \sum_{|\alpha| \le k} |D^\alpha u(x)|^2 dx .$$

Similarly, by $\mathring{\mathcal{H}}_{(k)}(H_0)$ we denote the space of all $u \in \mathcal{H}_{(k)}(H_0)$ which can be extended to a function $U \in \mathcal{H}_{(k)}(R_0^n)$ such that $\operatorname{supp} U \subset \{x' \in R_0^n : x_0 \geqslant 0\}$, and the norm is given by

$$\|u\|_{(k)}^2 = \int_{H_0} \sum_{|\alpha'| \le k} |D^{\alpha'} u(x')|^2 dx' .$$

Finally $\mathring{\mathcal{B}}_{(k)}(H_+)$ denotes the completion of the space $\mathring{\mathcal{H}}_{(k+1)}(H_+)$ with respect to the norm

$$|||u|||_{(k)}^2 = \|u\|_{(k)}^2 + \sum_{l=0}^{k} \|D_n^1 u(\cdot,0)\|_{(k-l)}^2 .$$

The following theorem concerns the mixed problem in H_+ with homogeneous Cauchy conditions.

Theorem 7. Let P, Q_j, $j = 1,\ldots,\varkappa$, and k satisfy the conditions of Theorem 5. Then for each $f \in \mathring{\mathcal{H}}_{(k)}(H_+)$, $g_j \in \mathcal{H}_{(m-1+k-m_j)}(H_o)$, $j = 1,\ldots,\varkappa$, there exists exactly one solution $u \in \mathring{\mathcal{B}}_{(m-1+k)}(H_+)$ of the problem $Pu = f$, $Q_j u(\cdot,0) = g_j$.

The following inequality holds

$$|||u|||_{(m-1+k)} \leq K\left(\|f\|^2_{(k)} + \sum_{j=1}^{\varkappa}\|g_j\|^2_{(m-1+k-m_j)}\right),$$

where K is a constant independent of u, f and g_j.

Proof. The functions f and g_j may be extended to functions $f_1 \in \mathcal{H}_{(k,o),\eta}(E^{n+1}_+)$, $g_{1,j} \in \mathcal{H}_{(m-1+-m_j),\eta}(R^n_o)$, for some η, so that $f_1 = 0$ and $g_{1,j} = 0$ for $x_o < 0$. By Theorem 6 and the Corollary following it we deduce that there is exactly one solution of the equations, $u_1 \in \mathcal{B}_{(m-1+k,o),\eta}(E^{n+1}_+)$ which is equal to 0 for $x_o < 0$. Also it follows that u_1 does not depend on the way we extend f and g_j.
The required solution is obtained as the restriction of u_1 to H_+.

Remark. In the above theorem we can take any interval $T_1 < x_o < T_2$ instead of $0 < x_o < T$. By the theorems from first section it follows that the uniqueness of the solution is assured in larger classes of functions, for example in $\mathring{\mathcal{H}}_{(m-1+k)}(H_+)$.

The aforegoing theorems and the theory of the Cauchy problem enable us to give some results on the mixed problem for some domains in manifolds. We give one of the possible formulations.

Let M be a n+1-dimensional smooth manifold. Let $\varphi, \psi \in C^\infty(M)$ be two real functions such that $\operatorname{grad}\varphi(x) \neq 0$ for $x \in M$, and covectors $\operatorname{grad}\varphi(x)$, $\operatorname{grad}\psi(x)$ are linearly independent if $\psi(x) = 0$. Let P be a differential operator of order m on the manifold M, strongly hyperbolic with respect to $\operatorname{grad}\varphi(x)$, with smooth coefficients.

Moreover, let us assume that for each $x^0 \in M$ such that $\varphi(x^0) > 0$ there exist an open subset $G_{x^0} \subset M$ and a real function $\varphi_{x^0} \in C^\infty(G_{x^0})$ with the following properties:

$\operatorname{grad}\varphi_{x^0}(x) \neq 0$ for $x \in G_{x^0}$, the operator P is strongly hyperbolic with respect to $\operatorname{grad}\varphi_{x^0}(x)$ and the set $\{x \in G_{x^0} : \varphi(x) \geq 0 , \varphi_{x^0}(x) \leq \varphi_{x^0}(x^0)\}$ is a compact subset of G_{x^0}. Let us introduce the following notations:

$$\Omega = \{x \in M : 0 \leq \varphi(x) \leq T , \psi(x) \geq 0\},$$

ω_l, $l = 1,\dots,N$ - all components of the set $\{x \in \Omega : \psi(x) = 0\}$,

$\omega_0 = \{x \in \Omega : \varphi(x) = 0\}$,

$$\tau_0(x) = \operatorname{grad}\varphi(x) , \quad \nu(x') = \operatorname{grad}\psi(x') \quad \text{for } x' \in \bigcup \omega_l ,$$

$T''_{x'}$ - any n-1-dimensional linear subspace of $T^{*}_{x'}$ complementing the subspace spanned by covectors $\tau_0(x')$ and $\nu(x')$. The components ω_l are n-dimensional smooth submanifolds of Ω. We assume that at each point they are not characteristic with respect to the operator P, i.e.

$$P^0(x', \nu(x')) \neq 0 \quad \text{for} \quad x' \in \bigcup \omega_l .$$

At each point x' we have the equality

$$P^0\left(x', \zeta_0 \tau_0(x') + \tau'' + \zeta_n \nu(x')\right) =$$

$$= P^0(x', \nu(x'))Q^+(x', \zeta_0, \tau'', \zeta_n) \cdot Q^-(x', \zeta_0, \tau'', \zeta_n) ,$$

where $\zeta_0 = \xi_0 + i\eta_0$, $\eta_0 < 0$, $\tau'' \in T''_{x'}$, Q^+ and Q^- are polynomials in ζ_n with roots in the upper and lower half-planes respectively. The degree $\varkappa_1$ of the polynomial Q^+ is constant in each ω_l. On ω_l there are defined $\varkappa_l$ differential operators $Q_{j,l}$ of orders $m_{j,l}$, $j = 1, \ldots, \varkappa_l$, with smooth coefficients.

If $N = \infty$, we assume that for some nonnegative integer k_0 $m_{j,l} \leq m-1+k_0$ for all j and l.

Theorem 8. Suppose that:

(A) for $x' \in \bigcup \omega_l$, $\xi_0 \in R^1$, $\tau'' \in T''_{x'}$, $\xi_0 \tau_0(x') + \tau'' \neq 0$ the real roots of the polynomial $P^0(x', \xi_0 \tau_0(x') + \tau'' + \zeta_n \nu(x'))$ in ζ_n are at most of double multiplicity;

(B) the polynomials $Q^0_{j,l}(x', \zeta_0 \tau_0(x') + \tau'' + \zeta_n \nu(x'))$ are linearly independent mod $Q^+(x', \zeta_0, \tau'', \zeta_n)$ for $x' \in \bigcup \omega_l$, $\eta_0 \leq 0$,

$\tau'' \in T''_{x'}$, $\zeta_0 \tau_0(x') + \tau'' \neq 0$.

Then for each $f \in \mathcal{H}^{loc}_{(k+1)}(\Omega)$, $u_0 \in \mathcal{H}^{loc}_{(m+k+1)}(\Omega)$, where $k \geq k_0$ and $D^j_{\tau_0}(f - Pu_0)\big|_{\omega_0} = 0$ for $j = 0,1,\ldots,k$, there exists exactly one solution $u \in \mathcal{H}^{loc}_{(m-1+k)}(\Omega)$ of the problem

$Pu = f$, $D^j_{\tau_0}(u - u_0)\big|_{\omega_0} = 0$, $j = 0,1,\ldots,m-1$,

$Q_{j,l}(u - u_0)\big|_{\omega_l} = 0$, $j = 1,\ldots,\varkappa_l$, $l = 1,\ldots,N$.

Proof. We start with two lemmas, one on local uniqueness of solutions and the second on local extendability of solutions. They are interesting in their own.

It may be proved that the conditions (A) and (B) are invariant under a change of variables.

Lemma 1. Let $x^0 \in M$, Ω a neighbourhood of x^0; $\varphi, \psi \in C^\infty(\Omega)$, grad$\varphi(x^0)$ and grad$\psi(x^0)$ are linearly independent. Let operators P, Q_j, defined respectively on Ω and on hypersurface $\psi(x) = \psi(x^0)$, verify the assumptions of Theorem 8 at the point x^0, and moreover let P be strongly hyperbolic with respect to grad$\varphi(x)$. Denote

$\Omega^+ = \{x \in \Omega : \psi(x) > \psi(x^0)\}$, $\omega = \{x \in \Omega : \psi(x) = \psi(x^0)\}$.

If $u \in \mathcal{H}_{(m-1+k)}(\Omega^+)$, $Pu = 0$, $Q_j u\big|_\omega = 0$, $j = 1,\ldots,\varkappa$ and $u = 0$ when $\varphi(x) < \varphi(x^0)$, then there exists a neighbourhood Ω' of the point x^0 such that $u = 0$ in Ω'.

Proof. Taking Ω sufficiently small and choosing suitable coordinate system we get:

$\Omega = \Omega_\delta = \{x \in R^{n+1} : |x| < \delta\}$, $x^0 = 0$, $\varphi(x) = x_0$, $\psi(x) = x_n$.

Let us put $\varphi_\varepsilon(x) = x_0 - \varepsilon + \sum_{i=1}^{n} x_i^2$, $0 < \varepsilon < \frac{1}{4}\delta^2$.

If δ and ε are small, then P is strongly hyperbolic with respect to grad $\varphi_\varepsilon(x)$. Thus changing once again the coordinate system we obtain: $u \in \mathcal{H}_{(m-1+k)}(\{x \in \Omega_\delta : x_0 < 0, x_n > 0\})$, $u = 0$ when $x_0 < -\varepsilon + \sum_{i=1}^{n} x_i^2$, $Pu = 0$, $Q_j u(\cdot, 0) = 0$, $j = 1, \ldots, \varkappa$. Extend the function u on the whole set $H_+ = \{x \in R^{n+1} : -\delta < x_0 < 0, x_n > 0\}$ putting $u = 0$ beyond Ω_δ, and operators P, Q_j so that extensions satisfy the assumptions of Theorem 7. Hence, by this theorem, $u = 0$ in H_+. So in the old coordinate system $u = 0$ in Ω_δ for $x_0 < \varepsilon - \sum_{i=1}^{n} x_i^2$. This completes the proof.

Lemma 2. Let Ω, φ, ψ, P and Q_j be the same as in Lemma 1. If $u \in \mathcal{H}_{(m-1+k)}(\Omega^+ \cap \{x \in \Omega : \varphi(x) < \varphi(x^0)\})$, $w_0 \in \mathcal{H}_{(m+k)}(\Omega^+)$, $Pu = 0$, $Q_j(u - w_0)|_\omega = 0$, $j = 1, \ldots, \varkappa$, then there exists a neighbourhood Ω' of x^0 and $u_1 \in \mathcal{H}_{(m-1+k)}(\Omega')$ such that $Pu_1 = 0$, $Q_j(u_1 - w_0)|_\omega = 0$, $j = 1, \ldots, \varkappa$ and $u_1 = u$ in $\Omega' \cap \{x \in \Omega : \varphi(x) < \varphi(x^0)\}$.

Proof. Using the same change of variables as in the proof of Lemma 1 we arrive at the following situation :

$u \in \mathcal{H}_{(m-1+k)}\left(\left\{x \in \Omega_\delta : x_0 < -\varepsilon + \sum_{i=1}^{n} x_i^2,\ \ x_0 < 0,\right.\right.$
$\left.\left. x_n > 0\right\}\right)$ $Pu = 0$, $w_0 \in \mathcal{H}_{(m+k)}(\Omega_\delta^+)$, $Q_j(u-w_0)(x',0) = 0$.

Let $\chi \in C_0^\infty(\Omega_\delta)$ be such that $\chi = 1$ in a neighbourhood of $\left\{x \in R^{n+1} : -\varepsilon + \sum_{i=1}^{n} x_i^2 \leq x_0 \leq 0\right\}$. Then

$P\chi u \in \mathcal{H}_{(k)}\left(\left\{x \in \Omega_\delta : x_0 < -\varepsilon + \sum_{i=1}^{n} x_i^2,\ x_0 < 0,\ x_n > 0\right\}\right)$

and $P\chi u = 0$ in a neighbourhood of the sets

$\left\{x \in R^{n+1} : |x| = \delta,\ x_0 < 0,\ x_n > 0\right\}$ and

$\left\{x \in R^{n+1} : x_0 = -\varepsilon + \sum_{i=1}^{n} x_i^2,\ x_0 < 0,\ x_n > 0\right\}$. Putting f equal to $P\chi u$ on the domain of this function and equal to 0 in remaining points of the set $H_+ = \left\{x \in R^{n+1} : -\delta < x_0 < 0, x_n > 0\right\}$ we obtain $f \in \mathring{\mathcal{H}}_{(k)}(H_+)$.

Similarly $Q_j\chi u(\cdot,0) \in \mathcal{H}_{(m-1+k-m_j)}\left(\left\{x \in \Omega_\varepsilon : x_0 < -\varepsilon + \sum_{i=1}^{n} x_i^2,\right.\right.$ $\left.\left. x_0 < 0,\ x_n = 0\right\}\right)$, $Q_j\chi u(x',0) = 0$ in a neighbourhood of $\left\{x' \in R_0^n : |x'| = \delta,\ x_0 < 0\right\}$ and $Q_j\chi u(x',0) = Q_j w_0(x',0)$ in a neighbourhood of $\left\{x' \in R_0^n : x_0 = -\varepsilon + \sum_{i=1}^{n-1} x_i^2,\ x_0 < 0\right\}$. So putting $g_j(x') = Q_j\chi u(x',0)$ for x' from the domain of $Q_j\chi u(\cdot,0)$, $g_j(x') = Q_j w_0(x',0)$ for

$x' \in \left\{x' \in R_0^n : x_0 \geq -\varepsilon + \sum_{i=1}^{n-1} x_i^2,\ x_0 < 0\right\}$ and $g_j(x') = 0$

in the remaining points of the set $H_0 = \left\{x' \in R_0^n : -\delta < x_0 < 0\right\}$ we get that $g_j \in \mathring{\mathcal{H}}_{(m-1+k-m_j)}(H_0)$.

By Theorem 7 there exists a function $u_1 \in \mathring{\mathcal{B}}_{(m-1+k)}(H_+)$ such that $Pu_1 = f$, $Q_j u_1(\cdot,0) = g_j$, $j = 1,\dots,\varkappa$. Of course $u_1 = \chi u$ on the domain of χu, thus in particular $u_1 = u$ in a neighbourhood of $\{x \in R^{n+1} : x_0 = -\varepsilon + \sum_{i=1}^{n} x_i^2,\ x_0 < 0,\ x_n > 0\}$, $Pu_1 = 0$ in a neighbourhood of $\{x \in R^{n+1} : x_0 \geqslant -\varepsilon + \sum_{i=1}^{n} x_i^2, x_0 < 0, x_n > 0\}$ and $Q_j(u_1 - w_0)(x',0) = 0$ in a neighbourhood of $\{x' \in R_0^n : x_0 \geqslant -\varepsilon + \sum_{i=1}^{n-1} x_i^2,\ x_0 < 0\}$. Hence we obtain the conclusion of Lemma 2.

Now we are ready to prove Theorem 8.

At first we are going to prove the uniqueness of the solutions. Let $u \in \mathcal{H}^{loc}_{(m-1+k)}(\Omega)$, $Pu = 0$, $D^j_{\tau_0} u|_{\omega_0} = 0$, $j = 0,1,\dots,m-1$, $Q_{j,l}u|_{\omega_l} = 0$, $j = 1,\dots,\varkappa_l$, $l = 1,\dots,N$. It is enough to prove that for each $x^0 \in \Omega$ $u = 0$ in a neighbourhood of this point.

By the properties of the function φ_{x^0} the proof of uniqueness reduces to the following statement : if G_0 is an open subset of M such that $\overline{G_0}$ is compact, $\varphi_0 \in C^\infty(G_0)$ is a real function such that P is hyperbolic with respect to grad $\varphi_0(x)$ for $x \in G_0$ and the set $\{x \in G_0 : \varphi(x) \geqslant 0, \varphi_0(x) \leq c_0\}$ is a compact subset of G_0 for some c_0, then $u = 0$ in the set $\{x \in G_0 : 0 \leq \varphi(x) \leq T,\ \psi(x) \geq 0,\ \varphi_0(x) \leq c_0\}$.

Since the uniqueness of solutions for the Cauchy problem the function u may be extended to a function $u \in \mathcal{H}_{(m-1+k)}(\Omega_{\varepsilon_1})$, where $\Omega_{\varepsilon_1} = \{x \in G_0 : -\varepsilon_1 \le \varphi(x) \le T,\ \psi(x) \geqslant 0\}$, by $u = 0$ for $-\varepsilon_1 < \varphi(x) < 0$. If ε_1 is sufficiently small, then $\{x \in G_0 : -\varepsilon_1 \le \varphi(x),\ \varphi_0(x) \le c_0\}$ is a compact subset of G_0. Denote

$$\Omega_{\varepsilon_1,c} = \{x \in G_0 : -\varepsilon_1 \le \varphi(x) \le T,\ \psi(x) \geqslant 0,\ \varphi_0(x) \le c\}.$$

Observe that there exists c such that $u = 0$ in $\Omega_{\varepsilon_1,c}$. We shall prove that if $c < c_0$, then for some $c_1 > c$ $u = 0$ in $\Omega_{\varepsilon_1,c_1}$. Let $x^1 \in \Omega_{\varepsilon_1,c}$, $\varphi_0(x^1) = c$, then $u = 0$ in some neighbourhood of the point x^1. Indeed, it is obvious when $\varphi(x^1) < 0$, if $\psi(x^1) > 0$ it follows from the uniqueness of solutions for the Cauchy problem, if $x^1 \in \omega_l$ for some l then it follows from the Lemma 1. Since $\Omega_{\varepsilon_1,c} \cap \{x \in G_0 : \varphi_0(x) = c\}$ is compact and in some neighbourhood of it $u = 0$, for $c_1 > c$ and $c_1 - c$ sufficiently small $u = 0$ in $\Omega_{\varepsilon_1,c_1}$. Hence we deduce that $\sup\{c : u = 0 \text{ in } \Omega_{\varepsilon_1,c}\} = c_0$. This establishes uniqueness of solutions.

It remains to prove the existence of solutions. Extend the functions f and u_0 on the set $H_{[0,T]} = \{x \in M : 0 \le \varphi(x) \le T\}$ so that $f \in \mathcal{H}^{loc}_{(k+1)}(H_{[0,T]})$, $u_0 \in \mathcal{H}^{loc}_{(m+k+1)}(H_{[0,T]})$ and consider the solution of the Cauchy problem $Pv_0 = f$,

$$D^j_{\tau_0}(v_0 - u_0)\Big|_{\varphi(x)=0} = 0, \quad j = 0,1,\dots,m-1.$$

There exists exactly one solution of this problem $v_0 \in \mathcal{H}^{loc}_{(m+k)}(H_{[0,T]})$. By the equality $D^j_{\tau_0}(f - Pu_0)\big|_{\varphi(x)=0} = 0$, $j = 0,1,\ldots,k$ it follows that $D^j_{\tau_0}(v_0 - u_0)\big|_{\varphi(x)=0} = 0$ for $j = 0,1,\ldots,m+k$. Let w_0 be a function on $H_{[-\varepsilon_1,T]}$ defined by $w_0 = v_0 - u_0$ on $H_{[0,T]}$ and $w_0 = 0$ on $H_{[-\varepsilon_1,0]}$. We have $w_0 \in \mathcal{H}^{loc}_{(m+k)}(H_{[-\varepsilon_1,T]})$. If we show that there exists a function $v \in \mathcal{H}^{loc}_{(m-1+k)}(\Omega)$ such that $Pv = 0$, $D^j_{\tau_0} v\big|_{\omega_0} = 0$, $j = 0,1,\ldots,m-1$, $Q_{j,l}v\big|_{\omega_l} = Q_{j,l}w_0\big|_{\omega_l}$, $j = 1,\ldots,\varkappa_l$, $l = 1,\ldots,N$, then $u = v_0 - v$ is a solution of our problem.

At first we show that there exists a function v with the above properties but which is solution of these equations on the set $\{x \in \Omega : \varphi_{x^0}(x) < \varphi_{x^0}(x^0) + \varepsilon_0\}$ for ε_0 sufficiently small. Let $\Omega_{\varepsilon_1,c}$ be as before except that φ_0 is now replaced by φ_{x^0}, and let $c_0 = \varphi_{x^0}(x^0) + \varepsilon_0$. It is easily seen that for some c the solution of this problem in $\Omega_{\varepsilon_1,c}$ is given by $v = 0$. It follows from Lemma 2 and theory of Cauchy problem that a solution $v \in \mathcal{H}_{(m+k)}(\Omega_{\varepsilon_1,c})$ for $c < c_0$ of the equations may be extended on some neighbourhood of $\Omega_{\varepsilon_1,c}$ in Ω to a solution of these equations. Hence we infere that there exists $c_1 > c$ and $v_1 \in \mathcal{H}_{(m+k)}(\Omega_{\varepsilon_1,c_1})$ which is a solution of these equations. From this we deduce in the usual way that there exists a solution v defined on whole $\Omega_{\varepsilon_1, c_0}$.

Let $x^1, x^2 \in \Omega$ and let v_{x^1}, v_{x^2} be the corresponding solutions defined on the sets $\Omega^1_{\varepsilon_1, c_1}$ and $\Omega^2_{\varepsilon_2, c_2}$ respectively. Then v_{x^1} and v_{x^2} coincide on $\Omega^1_{\varepsilon_1, c_1} \cap \Omega^2_{\varepsilon_2, c_2}$.

Indeed, this set is equal to

$$\Omega_0 = \left\{ x \in \Omega : \varphi_{x^1}(x) \le c_1, \ \ \varphi_{x^2}(x) \le c_2 \right\},$$

and putting $\varphi_0(x) = (\varphi_{x^1}(x) - c_1)(\varphi_{x^2}(x) - c_2)$ for $x \in \Omega_0$ we get that P is strongly hyperbolic with respect to grad $\varphi_0(x)$. Because $v_{x^1} - v_{x^2}$ is a solution of the homogeneous equations in Ω_0, by already proved uniqueness of solutions we get that $v_{x^1} = v_{x^2}$ in Ω_0.

Thus the family of all functions v_x corresponding to the points $x \in \Omega$ define a solution v in Ω. This completes the proof of Theorem 8.

Remark. It is seen from the proof that if we use the spaces $\mathcal{H}^{loc}_{(s)}(\Omega)$ with arbitrary s, then the assumptions on regularity of f and u_0 may be weakened.

The assumption of existence of the functions φ_{x^0} is satisfied for example for a uniformly strongly hyperbolic operator with respect to x_0, defined on R^{n+1}.

4. Concluding remarks.

We end this paper with a short discussion of the obtained results.

There are distinguished two problems: conditions under which the inequality (0.1) or (1.5) hold, and conditions assuring the correctness of the mixed problem. If the first problem is concern then it seems to the author that the conditions (A) and (B) may not be improved in essential way. It may be proved that (B) is necessary condition for (1.5), and examples shows that (A) is also rather essential condition.

In the case of equations of the second order the condition (A) is always verified and thus (B) is necessary and sufficient condition for (1.5). The condition (B) is strongly restrictive, for instance it is not verified for Neumann problem for the wave equation. Therefore a progress in the theory of mixed problem is connected with inequalities weaker then (1.5).

The methods used in this paper may be applied to more general mixed problems than here.

The above theory may be overcarried on strongly hyperbolic systems of equations satisfying conditions analogous to (A) and (B). But obtained theory is not so interesting because these conditions are too much restrictive. (1.5) holds for much more general systems of equations.

Thus it is seen that the obtained here results may be treated as an introductory step, especially in the theory of systems of equations.

It is worth of mentioning that there are a lot of other interesting problems connected with the mixed problem, e.g. asymptotic methods, reflection, singularity of solution e.t.c. These problems were not concerned in this paper and there is known very little about them.

The author is greatly indebted to Professor B. Bojarski for very helpful comments.

References

[1] S.Agmon, Problèmes mixtes pour les équations hyperboliques d'ordre supérieur in Les équations aux dérivées partielles, Coll.Int.du C.N.R.S., Paris, 25-30 Juin, 1962, pp.13-18.

[2] T.Bałaban, On the mixed problem for a hyperbolic equation, Bull.de l'Acad.Pol.des Sc., Série des sc.math., astr. et phys., Vol.XVII, No.4, 1969, pp.231-235.

[3] L.Gårding, Cauchy's problem for hyperbolic equations, Winter and Spring Quarters, 1957, University of Chicago.

[4] L.Hörmander, Linear partial differential operators, Springer, Berlin, 1963.

[5] M.Ikawa, Mixed problems for hyperbolic equations of second order, J.Math.Soc.Japan, 20, 1968, pp.580-608.

[6] M.Ikawa, A mixed problem for hyperbolic equations of second order with a first order derivetive boundary condition, preprint, Osaka University.

[7] J.J.Kohn, L.Nirenberg; An algebra of pseudo-differential operators, Comm.Pure and Appl.Math., 18, No.1/2, 1965, pp.269-305.

[8] G.Peyser, Energy integrals for mixed problem in hyperbolic partial differential equations of higher order, J.of Math. and Mech., Vol.6, 1957, pp.641-653.

[9] V.Thomée, Estimates of the Friedrichs-Levy type for mixed problems in the theory of linear hyperbolic differential equations in two independent variables, Math.Scand.5, 1957, pp. 93-113.